Fruits, Vegetables, & Berries

Fruits, Vegetables, & Berries

An Arranger's Guide

Kally Ellis and Ercole Moroni

Chilton Book Company
Radnor, Pennsylvania

Publisher's Note
The author and publishers have made every effort to ensure that all instructions given in this book are safe and accurate, but they cannot accept liability for any resulting injury, loss, or damage to either property or person whether direct or consequential and howsoever arising. They would like to draw particular attention to the following:

Several of the arrangements in this book include candles. It is advisable to spray the arrangement with flame retardant and to make sure that all foliage is well clear of the candles. Never leave lit candles unattended.

Small berries and fruits should be firmly glued into position and kept out of reach of small children.

Do not touch your eyes after handling dry floral foam as it is an irritant.

A QUARTO BOOK

ISBN 0-8019-8761-X

A CIP record for this book is available from the Library of Congress

This book was designed and produced by
Quarto Publishing plc
The Old Brewery
6 Blundell Street
London N7 9BH

Senior editor Michelle Pickering
Senior art editor Elizabeth Healey
Editors Jane Donovan, Heather Magrill, Miranda Stonor
Designer Tania Field
Chapter openers & collages Sallyann Bradnam
Photographers Paul Forrester, Laura Wickenden
Prop buyer Susannah Jayes
Picture manager Giulia Hetherington
Art director Moira Clinch
Editorial director Mark Dartford

Typeset by Type Technique,
121A Cleveland Street, London W1
Manufactured by Bright Arts
(Singapore) Pte Ltd
Printed in Singapore by
Star Standard Industries Private Ltd.

Contents

Introduction

It is funny how attitudes can change over the years. Not so long ago most people seemed to have an inherent aversion to anything deemed to be "unnatural." From flowers to food and fashion, anything that was not derived directly from the earth was likely to be viewed with some scepticism, if not positive distaste. And we are the first to admit that we shared this view.

But, like time, technology marches on. In the modern world of floristry and flowers, a new wave of artificial materials has arrived, bringing products so life-like that they defy the very word "artificial." Cleverly constructed and finely manufactured to the highest standards, little difference can be discerned between the genuine article and the fake without the benefit of a close examination. Thanks to new plastics, new polymers, and new polyurethane finishes, even Adam could be tempted with a fake apple!

Up till now, you have probably never dreamed of using artificial materials; you imagine them to be cheap and distasteful. Unfortunately, poor quality material is still available, but overcome your prejudices and hunt out the gems. That is what this book is all about: not what is to be avoided – that should be only too obvious – but what you can achieve with a little expert guidance. Remember, going artificial does not necessarily mean that you are taking an environmentally unfriendly option. Many of the materials in this book are still derived from "natural" sources.

If you need further convincing, let's talk practicalities. Natural materials have only a limited lifespan. In time, they will wither, losing color and moisture. If you want your arrangement to withstand the ravages of time and the elements, then the time has come to fake it! Using artificial materials allows you to create a truly permanent display, one that is easy to clean and always beautiful to behold. And even if you do tire of your arrangement, you can simply dismantle it and use your materials over and over again.

Before you get started though, we would like to give you a word of advice. When you are looking at artificial materials, always let your eye be your guide. Do not compare fake fruits with the real thing – viewed close-up, they will always look inferior. But do compare like with like. Consider whether materials, when used constructively and imaginatively, will blend and merge successfully so that no one will actually realise you are "cheating." Do not be too stern in your examination of materials – remember that your toughest critics will not be quite so eagle-eyed as you are.

We have tried to use artificial materials whenever possible in this book. In some instances, though, we have introduced natural elements where the artificial alternative was simply not good enough. So the Halloween gourds are real, not plastic, and the festive foliages are fresh from the forest, not from the factory. However, it can only be a matter of time before satisfactory artificial alternatives are available – watch out for them.

Finally, you may be surprised at the lack of flowers in a book written by two florists. As the fine art of floristry continues to develop and evolve, so the boundaries are pushed farther and farther back. The displays we create, both in this book and in our shop, encompass all manner of unusual, interesting, and, occasionally, even mundane objects! There are no rules, just as there are no color codes. Experiment, initiate, dare to be different – you will be well rewarded for your effort!

Kally Ellis and Ercole Moroni.

Fruits, Vegetables, and Berries

Until recent times, floristry has traditionally been restricted to flowers and foliage alone and, on reflection, you would be forgiven for not associating fruits, vegetables, and berries with floral design. Nowadays, it is not unusual to see flowers, fruits, and vegetables used in conjunction with one another, but rarely have we witnessed a totally "edible" display.

Designing arrangements purely from artificial fruits, vegetables, and berries is a new concept, which is both exciting for and inspirational to the keen florist and home arranger. The wide range of new materials now available opens up fresh opportunities for being creative and daring. A little imagination and some practical know-how will go a long way when you are experimenting with a variety of forms and colors.

It has become remarkably easy these days to find authentic-looking specimens of fruits, vegetables, and berries. Manufacturers have made great strides in perfecting that "realistic look," although some examples are almost defiantly fake. Although you should, wherever possible, select those materials which appear more natural, do not dismiss those which are not – in the right environment, they can add a touch of humor or even quirkiness to a display. Whatever your choice of materials, it is important to make sure that you always have a good selection of forms, colors, and textures to achieve a pleasing result.

Most of the fruits, berries, and vegetables used throughout this book are designed to fool even the experts. The manufacturer has striven to produce materials that are as natural-looking as possible. Each finished product, whether it is an apple, tomato, or chili pepper, has undergone a coloring process involving the subtle use of thousands of shades of the same color to achieve that crisp, authentic appearance. Not only is the ideal color attained, but also the perfect shape and weight, so that the materials feel as well as look natural. Even the texture is life-like, with a surface that gives a little when squeezed and then firms up again when released, in the manner of real fruits and vegetables.

The composition of most of these fruits, berries, and vegetables consists of a foam base coated with a flexible covering. The marriage of these two ingredients makes for a product that is vibrant in color and practically indestructible, even in harsh weather. The true-to-life weight is achieved with a central filling of sand – a discovery we made when attempting to push a wire through the center of one of the larger-headed fruits!

Artificial materials are, however, more commonly and less expensively made from polystyrene, plastic, paper,

wood, latex, and polyester. Each of these materials has its own advantages. Lettuces and red cabbages are remarkably life-like when made from paper and sympathetically colored. Generally, paper vegetables are light in weight and, therefore, easy to use in large arrangements, garlands, or delicate work. Plastic and polystyrene materials can have either a matt or glossy finish, and, although less authentic in appearance, provide an element of fun and sometimes give humor to an arrangement.

Cost depends as much on the size of the object and whether it is stemmed or not as on the quality of the materials. Vegetables are largely sold individually and unstemmed, whereas fruits and berries are more frequently available in stemmed clusters. Artificial fruits and vegetables that have a ready-made stalk made from wire or wood make arranging that much easier. However, do not worry if the stem is non-existent; creating your own stems is really quite simple (see Techniques, page 26). When cutting a stem of fruits or vegetables into smaller parts, never discard any of the remaining foliage on the stem, as this can be used as an excellent filler.

You will notice that even the cheaper artificial materials are more expensive than the real thing. However, they are a good investment since you can use them time and time again, especially if you keep them clean and uncrushed. While they will not rot or drop pollen, these materials still need proper attention. Regular dusting and wiping over with a damp cloth will keep them fresh-looking and pristine.

Finally, it is important to remember that exactly the same basic arranging rules apply when working with artificial fruits and vegetables, as with the real thing (or even flowers). To be truly convincing, arrange artificial materials just as you would real ones, grouping the different shapes and colors, and getting a good balance with flowing lines. Proportion and balance, both visual and actual, apply in just the same way and require the same considerations. When selecting the items that you wish to introduce into a display, always work in odd numbers (i.e. at least three stems or pieces of one particular product). This will ensure a good balance of material.

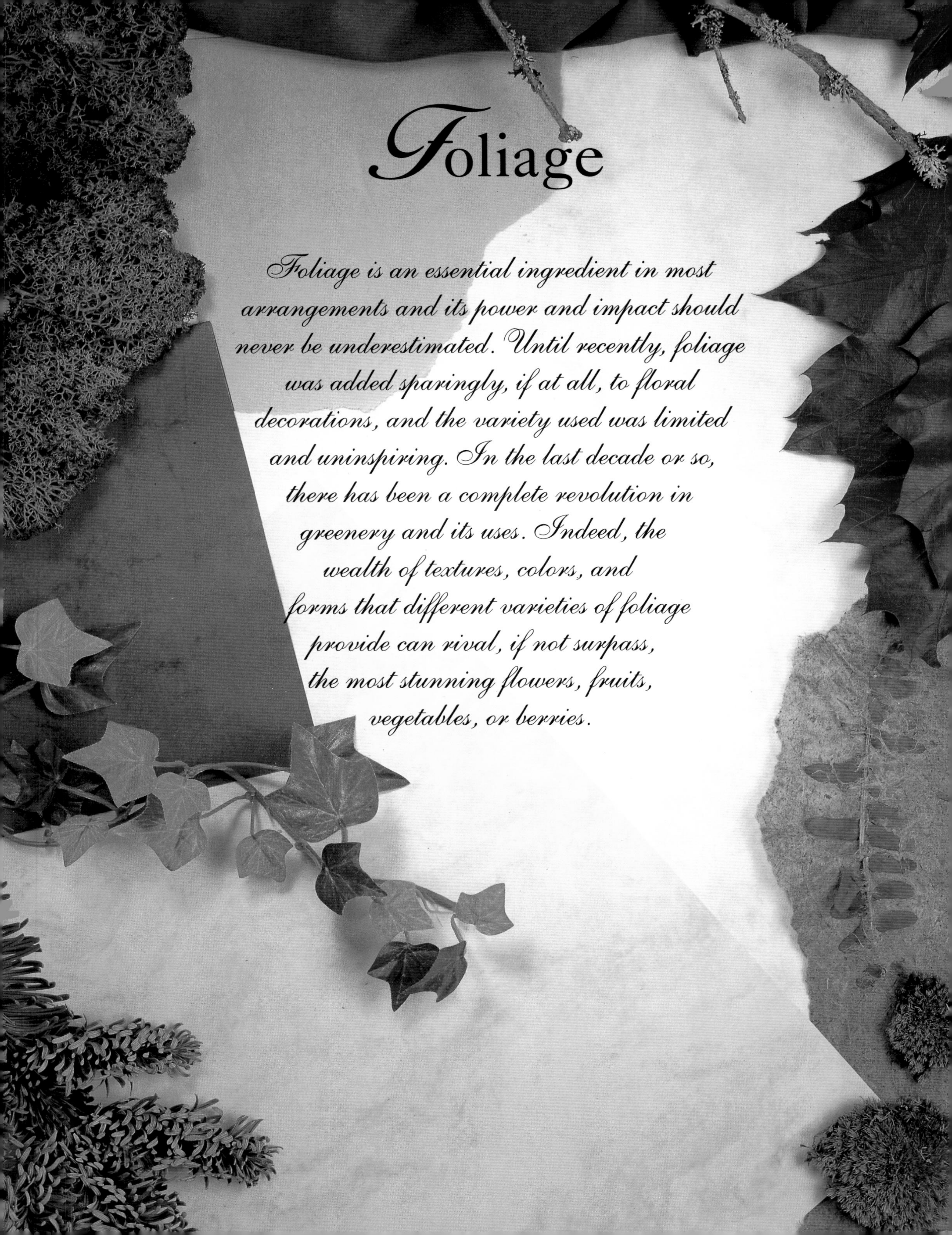

Foliage

Foliage is an essential ingredient in most arrangements and its power and impact should never be underestimated. Until recently, foliage was added sparingly, if at all, to floral decorations, and the variety used was limited and uninspiring. In the last decade or so, there has been a complete revolution in greenery and its uses. Indeed, the wealth of textures, colors, and forms that different varieties of foliage provide can rival, if not surpass, the most stunning flowers, fruits, vegetables, or berries.

COLOR AND TEXTURE

Foliage covers a wide spectrum of color and texture. We generally associate foliage with the color green. Although this is true in many cases, the varying shades and tints of green are endless, ranging from the dark, glossy holly leaf to the silvery, matt gray leaf of the eucalyptus tree. The color of foliage does not just stop at green – for instance, consider the rich, copper color of beech leaves and the orangey rusts of dried oak leaves.

Foliage takes on another dimension when it comes to bare branches, such as birch twigs, dogwood, lichen-covered larch, contorted willow, and pussy willow. These varieties offer endless possibilities to the inspired arranger, particularly in large displays where you want to add interest and create an unusual texture.

CHOOSING FOLIAGE

When choosing foliage, remember that most fruits, vegetables, and berries have their own. These can either be left intact and used in conjunction with the fruit, or removed and used in another display. Many fruits, vegetables, and berries, for example vine garlands, have a profuse amount of foliage so it is not always necessary to add extra greenery. Never discard a leaf, always save it, as you may find it will be useful for another arrangement.

ARTIFICIAL FOLIAGE

As with fresh foliage, the range of artificial foliage available on the market today is breathtaking. The most realistic-looking specimens are the glycerined variety, in particular eucalyptus, beech, oak, and magnolia. Once glycerined, the color of the foliage will change slightly to tones of khaki or shades of brown. This

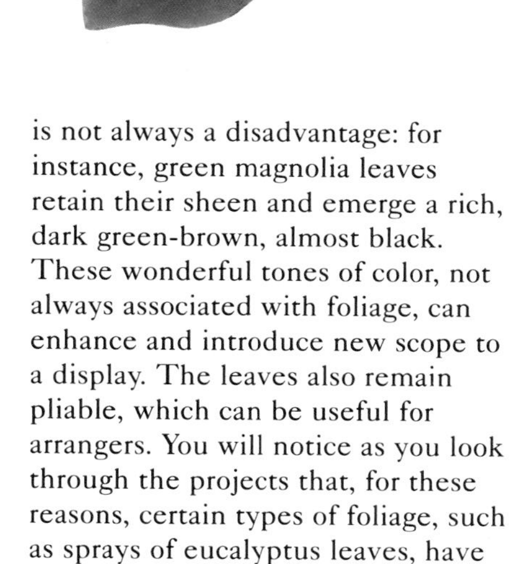

is not always a disadvantage: for instance, green magnolia leaves retain their sheen and emerge a rich, dark green-brown, almost black. These wonderful tones of color, not always associated with foliage, can enhance and introduce new scope to a display. The leaves also remain pliable, which can be useful for arrangers. You will notice as you look through the projects that, for these reasons, certain types of foliage, such as sprays of eucalyptus leaves, have been used repeatedly.

Other common forms of artificial foliage are constructed from silk, plastic, latex, and polyester. Some look more authentic than others, so it pays to shop around before making your final choice.

Unlike fresh foliage, the artificial stems of synthetic materials can be manipulated into almost any shape or line to give either a semblance of reality, or to add a sense of fun to an arrangement. The leaves, especially those connected to a fruit or vegetable, are often crumpled and squashed when they are first purchased. Do not worry as most leaves are flexible and can be remolded with your hands and opened out into the required shape.

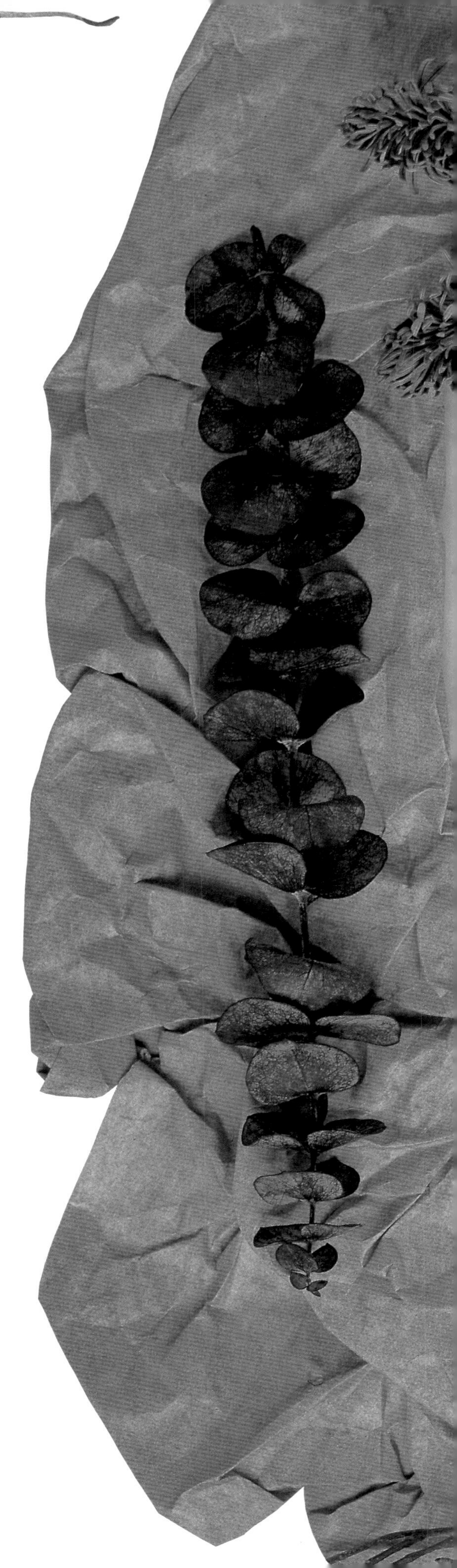

Accessories

Although the fruits, vegetables, and berries used throughout this book are visually appealing on their own, their appearance can be greatly enhanced by the addition of various other interesting materials which have a complementary and harmonizing effect on the arrangements. These accessories increase the creative scope of the arranger and give endless opportunities for being daring and adventurous.

The use of accessories allows more freedom of choice with regard to color, texture, and form, and gives rise to a multitude of arranging combinations. Do not be afraid to experiment: the results could be remarkably attractive and inventive.

Many of the featured accessories serve a dual purpose of being practical as well as decorative. Moss, for example, is widely employed to conceal the mechanics involved in wiring and to disguise floral foam. Cinnamon sticks, seed pods, fresh foliage, dried flowers, and candles can also be added to give extra interest and textural variety.

MOSS

There are several varieties of moss available, including bun moss, carpet moss, spaghnum moss, and reindeer moss. Whether you choose to buy these in fresh or dried form, the results are exactly the same. Each type of moss has its own attraction and versatility. You will notice that carpet moss, in particular, is used time and time again throughout the projects in this book, not only to conceal unsightly floral foam and wires, but also, with its lush, velvety green color and texture, as a highly decorative material.

Bun moss, with its dark green velvety texture, is perhaps our favorite variety of moss. It is available in clumps of various sizes and can be used in a similar way to carpet moss. The decoratively bulbous and furry appearance of this variety results in a more three-dimensional look.

Reindeer moss can be bought in almost any color. The species used in Hot Passion (see page 72) is silvery gray and perfectly complements the dark red of the chilies. Reindeer moss is usually treated with glycerine to guarantee long life and it is this treatment that makes it slightly rubbery to the touch. It is ideal for delicate work and bears no resemblance to bun or carpet moss.

Spaghnum moss is the least attractive of all the mosses, but it serves a practical purpose in providing a solid base for wreath frames, such as the Christmas Wreath on page 56. It is the most inexpensive form of moss you can buy and makes an ideal covering for floral foam, where it will not be regarded as a feature of the arrangement.

Whichever moss you choose, always remember to attach it securely to the floral foam using "hairpins" of wire. Any moss that is superfluous to requirements can be wrapped in a plastic bag and stored for many months in a cool location.

DRIED FLOWERS

Dried flowers are becoming increasingly popular as a form of decoration around the home. The marvelous colors and ever-increasing variety of materials available have made dried flowers an invaluable decorative asset, both on their own and alongside the fruits and vegetables in the projects that are to follow.

We have tried to limit the use of dried flowers in this book, a task which has been extremely difficult as they offer such a wide scope for inventiveness; also, the colors available are extraordinarily beautiful. The use of dried flowers in conjunction with fruits, vegetables, and berries creates an immediate impact. Dried red roses add an extravagance and vibrancy to the Heart-to-Heart project on page 68, while the bold sunflowers used in Autumn Sheaf (page 106) and Harvest Wreath (page 78) add warmth and interest, providing a strong focus for the arrangements.

RIBBONS

The range of ribbons that can be found is infinitely diverse: from silk, satin, and lace to hessian, rope, paper, and raffia. They come into their own as a final decorative touch to any arrangement. Wire-edged ribbon is ideal for making stylish bows as it can be twisted into almost any design and will keep its shape. Nowadays, more and more people are opting for the natural look and prefer rustic raffia, seagrass, and rope bows to finish off their displays.

Ribbons are not only highly decorative, they also have practical uses. Use them to disguise the mechanics involved in making a tied sheaf or Victorian posy, for example. Ribbon is also an excellent means of covering floral foam quickly and inexpensively, while simultaneously providing contrasting textural interest to an arrangement.

CANDLES

Candles are essential for creating atmosphere at that extra special dinner party or festive occasion; indeed, no dining-room table is complete without a little candlelight. Candles are usually a focal point in arrangements and therefore allow you to be economical with other materials. The Festive Centerpiece (see page 60), for example, is made almost entirely of festive foliages and very little else.

Nowadays there is an endless variety of candles to choose from, from plain, elegant church candles to brightly colored, twisted candles in shades of cerise pink and bright orange. Tealights can also be used in decorative displays and are available in diverse forms. The tealights disguised inside the top of the logs in the Natural Landscape (see page 40) are particularly novel and effective.

Remember, however, that candles can be dangerous if left unattended, particularly when you are using artificial or dried plant materials that may be highly inflammable. Always extinguish a candle whenever you leave a room.

FRESH FOLIAGE

We have greatly restricted the use of fresh foliage in this book. Indeed, the only projects to include it are those created for Christmas celebrations. When using fresh plant material, you must always consider its life expectancy. For this reason, we have selected blue pine branches and holly sprigs as they are sure to last throughout the festive season, and, even when they are dry, their allure is still very appealing. In addition to this, no artificial material can ever replace the fresh aroma that is released from pine, which evokes the perfect mood for the occasion.

WOODY PLANT MATERIAL

The wide range of woody plant material encompasses everything from the tall and slender branches of birch twigs, dogwood, and lichen to the smaller, more solid-looking pieces of bark, driftwood, cinnamon sticks, and lotus heads. All these materials are of a woody nature in both color and texture, and they can be added to any design to enhance both the main plant material and to add interest and textural contrast. The creative possibilities of these materials are infinite. Their natural and rustic appearance can only result in making a display look more relaxed and informal, particularly useful when a display contains a lot of artificial plant materials. The neutral, earthy colors of woody plant materials will complement any arrangement, so do not be afraid to experiment with them wherever possible.

Cinnamon sticks and lotus heads work particularly well in garlands and wreaths, whereas birch twigs and dogwood are better suited to longer, more free-flowing designs. Birch twigs are particularly versatile as they are highly flexible and can be manipulated into all kinds of forms. Use them to create a wreath base, for example, or twist them around other flexible materials to add interest and textural contrast.

Containers

Containers play an important role in any display. More often than not, it is the container which provides the inspiration for and determines the direction of an arrangement – its size, shape, color, and form. The rich variety of containers available means that you should always be able to find the one most suitable to the display you wish to create. The golden rule is to bear in mind the relationship between the container and the display, and the room or environment in which it is to be placed. Aim for harmony between all the elements and you will not go far wrong.

THE RELATIONSHIP BETWEEN PLANT AND CONTAINER

Consider your plant material, its size, shape, and weight. Small and delicate arrangements, using chili peppers, cornstalks, or fine eucalyptus stems, for example, would suggest the use of porcelain, glass, or, at a stretch, lighter-weight terracotta containers. More robust plant material, such as eggplants, corncobs, and dried sunflowers, would benefit from a heavier-looking pot to give them substance – perhaps stone or cast-iron. Try to avoid mixing a light-colored container with dark plant material. This will only serve to break up the harmony between the two elements and draw the eye away from the arrangement.

With solid, non-transparent containers, you have the added benefit of being able to conceal the mechanics of your display by using floral foam or scrunched-up chicken wire within the vessel to support your plant materials.

Clear glass receptacles are not ideal for this form of floristry, as most of the materials used do not have natural stems. As a result, you have to create false stems using wire and stem tape. This process, more often than not, gives a tidy but not altogether natural finish. Should you have a particular preference for glass, one way around this problem is to line the inside of the container with moss, for example. Alternatively, use pebbles, shells, potpourri, dried pasta pieces, berries, dried fruit

slices, and so on ... the list is endless. Depending on the shape and size of the glass container, this lining technique may require you to place a smaller, similar shaped receptacle inside the original container to hold the lining securely in place against the glass surface.

Lining is also relevant when you are using an expensive vase or bowl that you wish to protect. In this instance, it would be worthwhile making up your arrangement in a smaller plastic container, and then placing this inside the more attractive outer vase – the effect will be exactly the same and you will not stain or damage your vase.

In this book we have used both artificial and dried fruits, berries, and vegetables as our basic raw materials. Therefore, it is worth bearing in mind that your container does not have to be waterproof. Porous containers, such as terracotta, stone, and basketware, are obvious alternatives to glass and ceramics.

Finally, it is very important to consider whether the style of your arrangement is nostalgic and relaxed or modern and sophisticated. It will all have a bearing on the relationship between plant and holder. There are no strict rules. What is pleasing to the eye is the best guide.

BALANCE BETWEEN CONTAINER AND ENVIRONMENT

The balance between the container and the room within which it is situated depends both on the style of room and your own personal taste. The size and decorative finish of the container must echo and harmonize with its surroundings, not make it a jarring element.

Shape is an important factor, simplicity often being key. Just as harmony between the container and the arrangement is used to the benefit of the plant material, the whole ensemble should enhance the room and be sympathetic to it. In our modern, design-conscious age, the preference is for a clean, spacious look to a room. A container with a verdigris or distressed surface is a perfect foil for a plain wall and complements the plant materials very well. Such a pot will also work in a more traditional room, although you must be careful not to allow the surface texture to clash with any patterned background surfaces. Weathered terracotta is another option to consider.

If you are unable to get hold of naturally distressed and weathered containers, try achieving the same look with different paint techniques. The appearance of a new or plain container can easily be transformed by using techniques such as crackle-glazing, antiquing, and marbling.

All in all, the guiding words for teaming a container to an environment are harmony and simplicity. Stick to these guidelines and you will not go too far wrong.

CONTAINERS

Do not feel that you have to use traditional containers such as glass, cut glass, and china vases. There is a whole range of alternative ideas to be explored. Terracotta, wood, palm-leaf, and bamboo pots, as well as utility ware, can be used. The look that you achieve is restricted only by your own creativity.

The first items to consider using are those around you. Ceramic jars, such as flour containers and cookie barrels, will always bring a rustic appeal to your floral designs. Old china teapots have possibilities. Copper kettles, saucepans, and cooking pots of all descriptions will sit quite happily in a more traditional setting, as will glass storage jars and Kilner or pickling jars. Both large and small jugs, whether they are patterned, glazed, or earthenware, complement dried arrangements.

Utility ware is an endless source of ideas. Watering cans, pails, and flower buckets are very useful. Scour your local hardware store and bric-à-brac merchants. Wooden barrels and crates can also be used to contain the dried floral foam that holds your arrangement. Wicker, rattan, wood, and even terracotta tiles, wired or tied together with seagrass or twine, will create exciting new looks. Embark on a textural adventure – it's well worth the journey!

MAKING YOUR OWN CONTAINER

With a little imagination and a few basic raw materials, making your own container can be easier than you think. For instance, take a simple plastic container and cover it with overlapping glycerined magnolia leaves. Next, tie an attractive ribbon, raffia, or seagrass bow around the middle to secure. Alternatively, cover the plastic container with double-sided adhesive tape and smother it with dried leaves and thin twigs.

Other interesting materials to use when creating your own container include tree bark, wood, and hessian,

all of which have unusual textures. One of our favorite ideas is to use carpet or bun moss against a framework of chicken wire, which can be manipulated into any shape you require, such as a heart. This is a very cost-effective way of creating something original, yet attractive, and as the container is a neutral color, it will complement any setting, whether indoors or outside.

Alternatively, large fruits and vegetables, such as melons and gourds, hollowed out and used as containers will add creative flair and give a natural look to your arrangements. Group them together for a bold display and use them to maximum effect in a large room.

As it is not necessary to use water in your containers, all arranging possibilities can be entertained. Papier mâché is worth considering if you are after something different – you can make an imitation ceramic vase, animal-shaped containers, or even paper fruit. Be as bold and daring with your arrangements as you are in your choice of materials.

Equipment

Before you start, you need to get together a few basic tools of the trade to help you achieve professional results. Your local florist will stock most of the equipment listed here, but for the more specialist equipment, such as a glue gun and wire cutters, try your local hardware store or a supplier of flower-arranging sundries.

FLORAL FOAM

There are two types of floral foam available: the green variety, which is designed to be soaked and used in fresh flower arranging, and the brown kind that is used for dried and artificial arrangements. You need a good supply of the brown variety (dry floral foam), since it is ideal for holding wired stems in place. It comes in a variety of shapes and sizes, including bricks, rings, balls, cones, and flat "designer" boards, which can be cut into any shape.

FLORISTS' TAPE

This specialist waterproof tape is extremely resilient and adhesive. It is primarily used to secure foam in baskets, plastic trays, and containers. Two different thicknesses are available – wide and narrow. The thickness required is determined by the scale of the arrangement. Florists' tape comes in green and white colors. It is better to use the dark green variety as it is less conspicuous and resembles the natural color of the stems more closely.

FLORISTS' KNIFE

This is an essential tool, which must be kept sharp at all times. It is ideal for cutting through and shaping floral foam. The knife can also be used to trim the ends of more delicate stems.

STEM TAPE

Available in many shades of green, stem tape – also commonly known as gutta-percha – is essential for concealing all the mechanics of wired work. Not only does it give a neater finish, but it also has the added advantage of resembling a natural-colored stem.

FLORISTS' SCISSORS

A basic, yet essential piece of equipment, florists' scissors are vital for cleanly cutting off the ends of all sorts of thin, plastic stems, dried materials, and glycerined foliages, etc. Do not attempt to use these scissors to cut wire as this will damage the blades. A good pair of sharp scissors is also needed for cutting florists' tape and ribbons.

WIRE CUTTERS

Wire cutters are absolutely essential for working with artificial materials, as most stems will have a thick wire running through the center, particularly the branches of berries. With other, larger fruits and vegetables, where the stems are non-existent, you will need to create your own false stem out of stub wires. Again, wire cutters will come into their own here for cutting stub wires to the required lengths. If you are unable to obtain a pair of wire cutters, a pair of pliers can be substituted.

GLUE GUN

A glue gun is probably the most expensive item you will need. It is, however, an essential piece of equipment, especially when working with artificial materials. You will find it is particularly useful for delicate work, such as attaching small pieces of foliage or dried mushrooms to a wicker wreath frame. A glue gun is also invaluable for wiring up small fruit and vegetable heads, especially nuts. Low-temperature glue guns are the most commonly used, as there is less chance of suffering from painful burns should they make contact with the skin. A glue gun is remarkably easy to use as the molten glue released when you press the trigger dries in seconds. Make sure you buy the correct size of glue stick to fit your glue gun.

WIRES

You will need two types of wire for the projects in this book: reel wire and stub wire. Reel wire is a continuous length of wire wrapped around a spool and is used to bind materials in position, for example around a wreath frame. Stub wires are individual lengths of wire and are used to create or lengthen false stems and to make wire pins to hold moss in place.

Both types of wire are available in various lengths, weights and gauges (thicknesses). These range from the very fine silver rose wire frequently used in bridal and other delicate arrangements, through to medium-gauge wire, which is useful for supporting medium-sized lightweight heads, and finally, heavy-gauge wire, which is necessary for supporting large and often weighty fruit and vegetable heads (particularly the artificial variety). You will probably discover during the course of making up arrangements that the heavy-gauge wires prove to be the most useful. When purchasing wires, you need to specify the required gauge in millimetres or grams, i.e. Heavy-gauge: 1.00 to 1.25 mm (24 to 26 g); Medium-gauge: 0.71 to 0.90 mm (20 to 22 g); Light-gauge: 0.46 to 0.56 mm (18 to 19 g).

CHICKEN WIRE

Chicken wire, sometimes known as wire netting, has a multitude of uses. It is most commonly used to give added support to stems, when it is crushed into a loose, ball shape, then placed inside a container. Chicken wire can also be used for holding several blocks of floral foam together, when constructing large arrangements or unusually shaped displays. It offers scope for a variety of creative possibilities, in that it can be manipulated into any shape to form a skeleton framework when constructing your own containers.

Techniques

There are some techniques, including wiring and gluing, specially used for working with fruits, vegetables, and berries. Easy to learn, these will ensure that your arrangements always look superb.

GLUING

In both dried and artificial floristry, a glue gun is an invaluable tool to have. The molten glue, which is released through the nozzle by pressing the trigger, dries in seconds. A glue gun is particularly useful for attaching small, light materials to swags and wreaths, for example fungi, eucalyptus sprigs, and reindeer moss. It is also perfect for attaching moss to chicken wire.

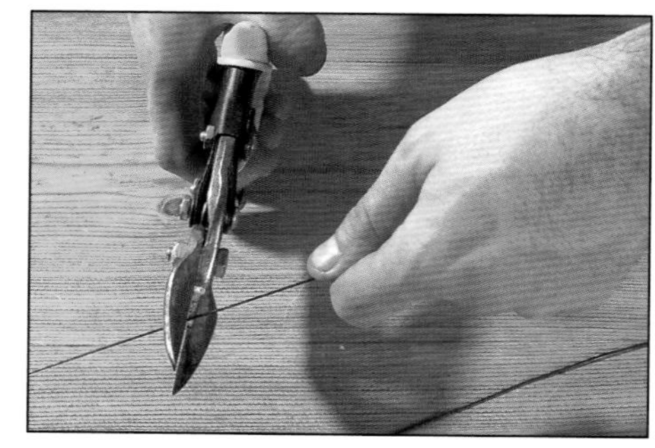

CUTTING WIRE

Most artificial foliage, fruit garlands, and clusters of fruit have wires running through their stems. Do not attempt to cut this wire with scissors as it will not only damage the blades but will require an enormous amount of effort. Instead, use a pair of wire cutters or even pliers to do the job.

SHAPING FLORAL FOAM

Floral foam comes mainly in brick-shaped blocks although other shapes and sizes are available, for example cones and spheres. The foam needs to be shaped to the required design. Each individual display is unique and will require its own amount of foam, which should be shaped accordingly. Using a sharp knife, trim off the square edges to give a more rounded look – this increases the surface area of the block and makes stem insertion easier.

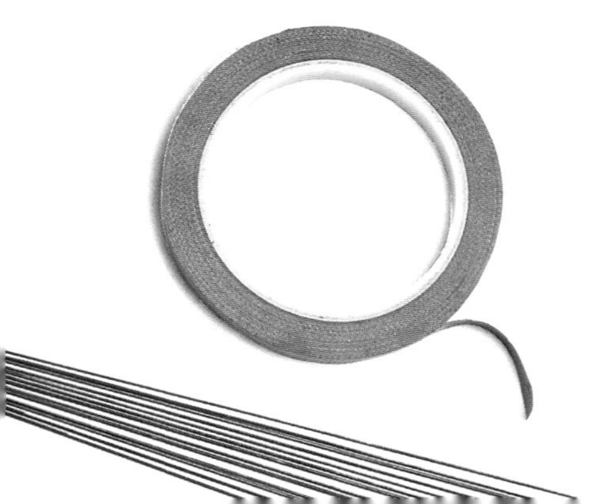

SECURING FOAM IN A BASKET

There are two ways of securing foam in a basket – with florists' tape or reel wire. In this instance we have used reel wire, but the principle is the same for both materials. Place your brick of floral foam in the center of the basket. Thread a length of wire through the twigs of the basket and loop it over the edge. Twist the wire around itself to secure. Stretch the wire diagonally across the basket, pressing it firmly down into the foam. Thread the wire through the twigs on the opposite side, pull taut, and twist to secure. Repeat this process to form a cross shape over the basket.

SECURING MOSS TO FOAM

To conceal the unsightly floral foam used in the mechanics of certain arrangements throughout the book, we have often used a layer of carpet moss. To secure the moss in position, make several "U"-shaped pins out of medium-gauge stub wire. Push the pins firmly through the moss and into the floral foam.

STEM WRAPPING

Stem wrapping, or guttering, with stem tape is an essential technique used in floristry, particularly where artificial materials are concerned. Not only does it add extra support to the stems, but it also makes them waterproof. In addition, as the color of the tape is green, it gives a semblance of reality to false stems. Once the technique is mastered, wires and stems can be covered in seconds. Hold the end of the stem tape on the neck of the fruit where it joins the false stem. Wrap the tape once or twice around the stem to secure and then hold the tape taut in one hand. Hold the false stem between the finger and thumb of your other hand and begin to twirl it, gradually moving the fruit away from you so that the tape travels down the length of the stem. The idea is to overlap the tape onto itself neatly as you work your way down the stem. Finish by twisting the tape onto itself to seal and tear off any excess.

CREATING FALSE STEMS

Most artificial fruits and vegetables are without stems of any kind. Therefore, to use them in an arrangement, a false stem has to be created. Depending on the weight of the fruit or vegetable, take the appropriate gauge of stub wire and thread approximately 3/8 in (1 cm) to 3/4 in (2 cm) through the base of the item. Pull both ends downwards to meet each other in a hairpin shape, and twist one tightly around the other to secure. Cover with stem tape to produce a natural look.

CREATING FALSE STEMS FOR HEAVY FRUITS AND VEGETABLES

Many artificial fruits and vegetables are heavy as they are filled with sand to give a semblance of the natural weight. Wiring such fruits and vegetables is quite tricky – you will need to apply quite a lot of pressure on the wire to pass it through the body of the fruit or vegetable.

1 Pass a thick-gauge stub wire horizontally through the base of the fruit. Turn the fruit round and repeat this process to form a cross.

2 Holding the fruit firmly, push all four ends of wire firmly downwards to meet each other.

3 Twist all the wires around one another to form a very sturdy and secure false stem. Cover with stem tape to produce a natural look.

LENGTHENING STEMS

This method is used where long stems are needed to form the outline shape of the arrangement, such as a shower bouquet. Take a length of thick-gauge stub wire and place it alongside the original false stem so that they overlap by approximately 2 in (5 cm). Bind the two wires tightly together with stem tape. For extra security, glue the two wires together before wrapping the stem. Once the whole length of the new elongated stem has been covered, twist and break off the tape.

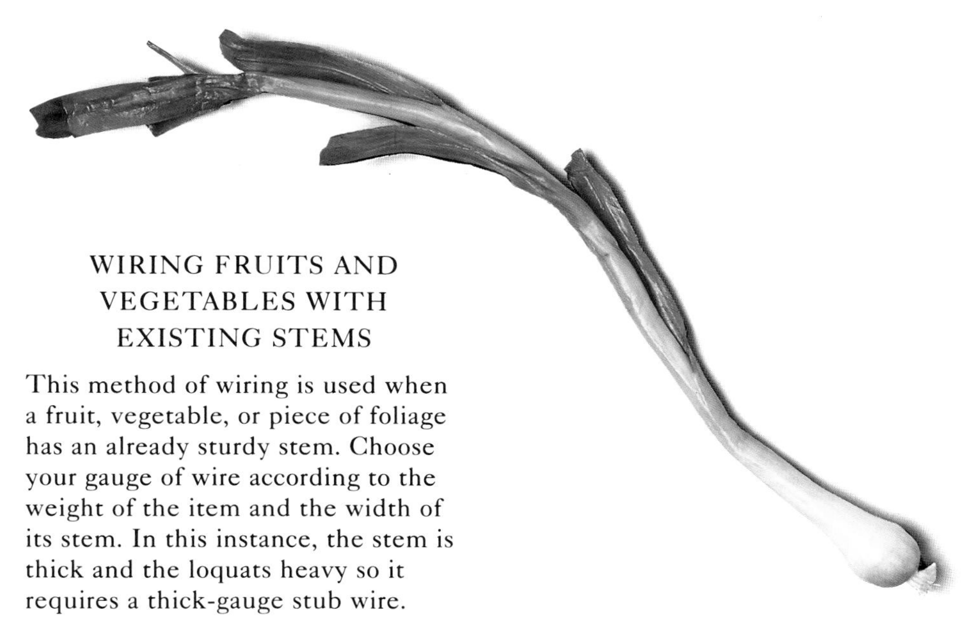

WIRING FRUITS AND VEGETABLES WITH EXISTING STEMS

This method of wiring is used when a fruit, vegetable, or piece of foliage has an already sturdy stem. Choose your gauge of wire according to the weight of the item and the width of its stem. In this instance, the stem is thick and the loquats heavy so it requires a thick-gauge stub wire.

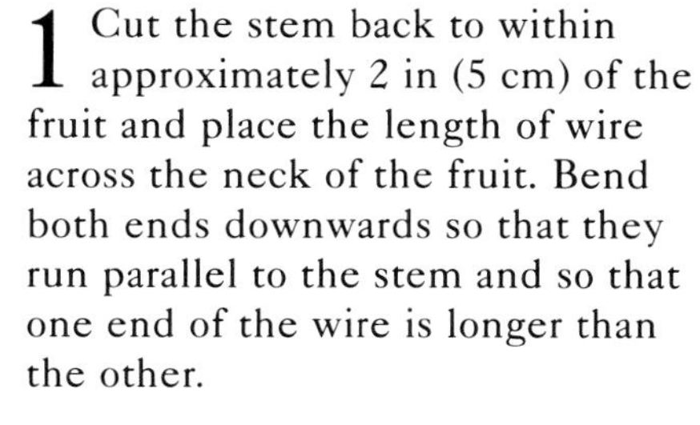

1 Cut the stem back to within approximately 2 in (5 cm) of the fruit and place the length of wire across the neck of the fruit. Bend both ends downwards so that they run parallel to the stem and so that one end of the wire is longer than the other.

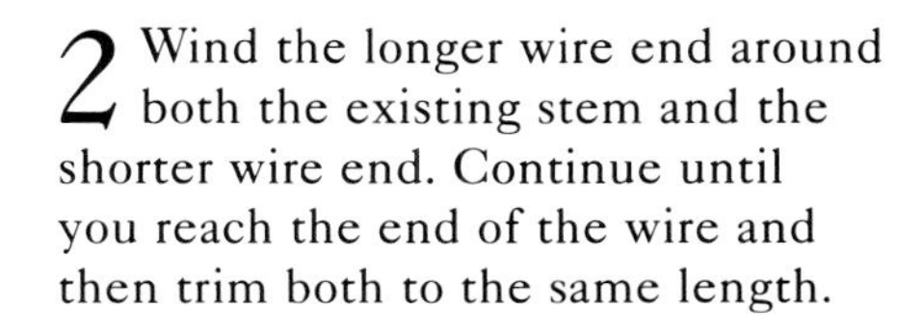

2 Wind the longer wire end around both the existing stem and the shorter wire end. Continue until you reach the end of the wire and then trim both to the same length.

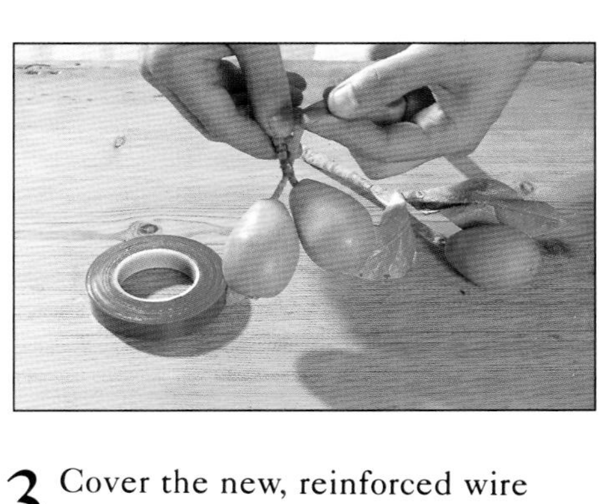

3 Cover the new, reinforced wire stem with stem tape to give a more natural look.

WIRING CONES

To wire pine cones, pass a length of medium-gauge stub wire through and around the prongs of the lower part of the cone until the two ends meet. Twist the wire ends around one another to create a false stem. Cover with stem tape, if required.

WIRING NUTS

Pass a length of medium-gauge wire as far as it will go through the natural hole at the base of the nut. Using a glue gun, release a drop of molton glue where the wire meets the nut. Hold the wire firmly in position until the glue sets. Wrap the wire with stem tape, if required.

WIRING SPONGE MUSHROOMS

Sponge mushrooms are one of the most attractive types of mushroom available. They can be used to add texture to an arrangement and to hide unsightly mechanics.

1 Using a pair of scissors, make two holes near the center of the base of the sponge mushroom. You will need to apply considerable pressure since the sponge is very hard.

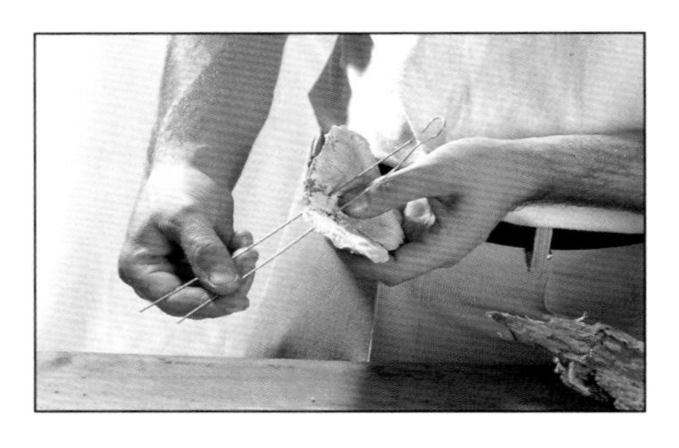

2 Make a hairpin shape out of a length of thick-gauge stub wire. Pass each end of the wire through the holes created by the scissors.

3 Turn the sponge mushroom over and twist the wire arms around each other, as close to the base of the mushroom as possible. Trim the false stem to the required length and cover with stem tape, if required.

SECURING CANDLES INTO ARRANGEMENTS

To wedge a candle into a foam base, you will need to make four "U"-shaped pins out of thick-gauge stub wire. Place the curved end of the wire against the candle about 1 in (2.5 cm) above the base. Secure by winding strong florists' tape tightly around the bottom of the candle and the curved ends of the wire pins several times. Insert the wire prongs firmly into the floral foam until the candle base is resting directly on top of the foam.

SECURING TERRACOTTA POTS INTO ARRANGEMENTS

Terracotta pots, particularly the old, handmade variety, have holes at the base. Depending on the size and weight of the pot, bend one or two lengths of thick-gauge stub wire in half. Place a short stick (no longer than the width of the pot base) horizontally in the bend of the pin and wrap the wire around the stick to hold it securely. From inside the pot, thread the two ends of wire through the holes at the base and pull them through until the stick is held flat. Twist the two wire ends together to form a false stem and cover with stem tape, if required.

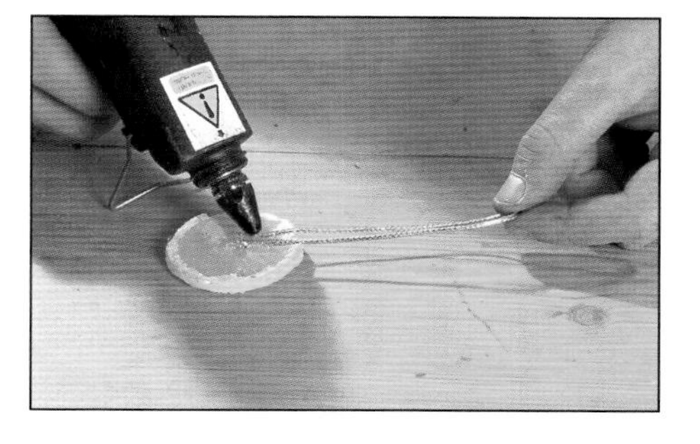

HANGING LOOPS

Although artificial fruit slices look very authentic, in reality they are totally inflexible and rigid. Trying to make a hole through which to thread a ribbon would be a futile exercise. Instead, cut a length of ribbon or string to the required size. Hold both ends against the fruit slice to form a loop. With a glue gun, release a couple of drops of hot glue onto the ends and hold in position until the glue has set.

MAKING A BOW

Bows are used to finish bouquets, decorate swags and garlands, and add interest to basket and table arrangements. The wire-edged ribbon used here can be twisted into any design and will retain its shape.

1 Fold the ribbon over to form two loops in the shape of a figure-of-eight. Pinch the fabric together at the cross-over point.

2 Holding the ribbon at the cross-over point, repeat the figure-of-eight to make four loops, again pinching the fabric together at the center. Cut the ribbon diagonally using a sharp pair of scissors.

3 Take a separate length of ribbon and use this to tie the bow at the cross-over point. To finish, trim off any excess ribbon and shape the loops with your fingers.

Drying and Preserving

While everyone appreciates the short-lived delight of fresh flowers, few people can afford the luxury of buying them on a regular basis. However, an ever-broader range of dried and preserved plant materials has become available over the last few years. Dried and preserved materials provide an enduring pleasure which can last for years, and, as a consequence, this can save you a great deal of money in the long run.

Both dried and preserved plant materials look so much fresher and prettier these days, due to the many different ways in which they are treated to give them longer life. There are various methods of preserving plant material and some are more suited to certain types of product than others. However, no single method has guaranteed results. Success depends not only on extracting all the moisture from the plants, but also on picking the material at the right time and at the correct temperature and humidity level.

It is important to note that the drying and preserving processes do not directly apply to the artificial fruits, vegetables, and berries used throughout this book as they are mainly composed of man-made fibers. However, these processes are relevant to certain arranging materials featured within these pages, such as foliage, grasses, dried flowers, artichokes, pomegranates, apples, and fruit slices, which are all invaluable to arrangers.

The two main preserving processes that directly relate to the plant materials used in this book are air-drying and glycerining.

AIR-DRYING

Air-drying is the simplest and most common means of preserving flowers and other types of plant material. It is best suited to drying materials such as maize, wheat, artichokes, fruit slices, pomegranates, moss, and lichen, to name but a few.

The main requirement for this process is a warm, dry area with plenty of circulating air. A dark place for storing materials during the drying process will also help preserve the color – a warm cupboard, for instance, would be ideal, or an attic, garden shed or garage. The temperature should be a minimum of 50°F.

Remember to strip all the lower leaves from the stems and tie them into manageable bunches. Rubber bands are better than string for holding the branches together since they continue to grip the stems in place while they dry out and shrink. Most materials can then be hung upside-down from nails or a drying rack. Large-headed materials, such as artichokes, pomegranates, and maize, can be laid out to dry on sheets of newspaper or chicken wire stretched over a box. Moss, fungi, and lichen can be dried in a similar fashion by loosely placing the clumps of material in a wooden fruit box or on crumpled sheets of newspaper. Space them well apart to allow air to circulate otherwise they will go moldy. When all the materials are papery to the touch, you will know that they are dry. This process can take anything from a week to a month, or even longer, depending on the type of material and the drying conditions.

GLYCERINING

This method is ideally suited to preserving foliage and reindeer moss. Like air-drying, the basic technique is quite simple, but the real benefit of glycerined foliage is that it retains its suppleness and flexibility. The types of foliage that respond best to this technique include eucalyptus, beech, oak, and magnolia leaves. All these varieties are used time and time again throughout the book.

To make the preserving solution, combine two measures of hot water with one measure of glycerine. Fill a tall container with this mixture to a depth of several inches. Cut all the stems at a slant and place a few of them into the solution. Check that all the leaves that are covered by the mixture are stripped off. The stems will absorb the mixture and become preserved in the process. Top up the container as required.

You will find that the colors of the foliage will change through the glycerining process. Most arrangers love the superbly rich tones of glycerined foliage – the khakis, browns, rusts, and maroons. However, if you do not find these colors pleasing, you can always blend a few drops of food coloring into the glycerine, which opens out all kinds of possibilities for unusual shades and variations. When all the leaves have undergone a change in color, you can remove the stems and use them in your arrangement.

Finally, a word of warning: this method of preserving can be expensive. Experiment with a few stems first and see just how much of the mixture is absorbed during the process. Remember, it is better to have a few perfectly glycerined stems than several half-finished ones.

Color and Design

As with flowers, the art of arranging fruits, vegetables, and berries is a decorative expression of a number of artistic skills. It combines an eye for color with the appreciation of the wide variety of textures and shapes available. You do not have to be an artist to possess a sense of style and proportion and a feeling for good design.

The way in which you arrange fruits, vegetables, berries, and foliage will reflect your own individual style, personality, and even the mood of the moment. Any design should therefore be a free expression of the arranger's stylistic views and feelings. No design is "incorrect," although some arrangements may have a better awareness of balance than others, but this is something which is remarkably easy to learn and to achieve.

Design determines the way in which we select and use materials and the manner in which every item is planned and organized to give the maximum visual appeal. The elements of design, which guide us in creating visually attractive work, are not rigid – they allow the freedom and flexibility to express our own style. Understanding the elements of design is therefore essential. These can be broken down into three categories: color, form, and texture.

COLOR

When choosing a color scheme, you should take into account the setting, that is the style of the room and furnishings, the background against which a display will be seen, and the overall effect you wish to create.

To help you understand the way in which colors work together, it will help if you refer to what is known as the color wheel. This is made up of several color groups, but the three basic ones are primary, secondary and tertiary colors. Primary colors include red, blue, and yellow. They cannot be created by mixing other colors and are the source of all other colors. Secondary colors include green, orange, and violet. They are created when two neighboring primary colors on the color wheel are mixed in equal proportions. Tertiary, or intermediate colors, are made by combining equal parts of a

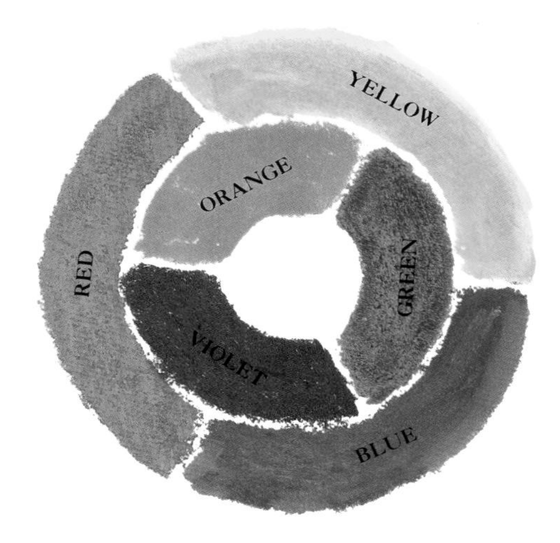

Reds, yellows, and oranges are warm, advancing colors.

Fruits and berries are ideal for rich, monochromatic designs.

Complementary colors such as greens and reds illustrate the attraction of color opposites.

Stunning results can be achieved by pairing contrasting cool and warm colors such as blue and yellow.

primary color and the secondary color that lies next to it.

With the aid of the color wheel, you can evaluate the effect each color has on its neighbor, and therefore the impact each differently colored vegetable, fruit, or grouping of materials would have on its companions in an arrangement. When two colors that oppose each other on the wheel are used together – primary blue and secondary orange, for example – the results can be stunning. These pairings, which illustrate the attraction of color opposites, are known as contrasting or complementary hues. At the other end of the scale are the color pairings termed "adjacent." These are positioned next to each other on the wheel and are harmonious, for example red, mauve, and blue. Frequently colors are described as "advancing" or "receding." Advancing colors are warm shades, such as reds, yellows, and oranges. Receding colors are cool hues, such as blues, greens, and violets.

It is widely considered that the most important aspect of color and design theory is a thorough knowledge of the receding and advancing colors, and how they affect people emotionally. Warm and cool colors move people in different ways and they always have an instant impact on moods and feelings. Individual preferences, for example, can draw you to warm colors when you want to feel uplifted, while cool colors soothe you.

Another popular color theme is the "one color" or "monochromatic" color scheme, incorporating a full range of tints, shades, and tones that are predominantly one color, for instance the red shades of strawberries, raspberries, plums, cherries, and cranberries. The end result can be quite breathtaking.

FORM

Ideally, when you are choosing the materials to create a particular arrangement, you should take their shapes and sizes into full account. The contrast created between differing forms will add interest and scope. Arrangement forms generally fall into one of three groupings – linear, round, or spray. Some styles, such as topiaries, require the use of only one while other designs may incorporate all three groupings.

When you are working with fruits, vegetables, and berries you are more limited in choice of form than you are with flowers, since most of the raw materials you will be using are rounded by nature. However, there are plenty of linear-shaped fruits and vegetables to experiment with, for example bananas, bunches of grapes, carrots, green onions, leeks, and Chinese lettuces. Tall, elongated materials, such as sprays of eucalyptus leaves, are generally used to give height and to break up solid, round shapes. The contrasting form between linear and round materials has a complementary effect on the overall design. Pointed shapes give movement and depth to a design. Foliage is also a good example of a linear material and this can be used to achieve the required height and width. The natural bend in the foliage branches ensures a sense of movement and softens the edges of any design.

Round materials, such as tomatoes, apples, potatoes, peaches, and so on, automatically attract the eye and are generally used as a design focus.

In addition to round and linear materials, there are transitional materials, which are sprays of fruits or berries on a stem. These act as a link between round and linear materials to blend the two together and to soften the overall effect. Spray foliage, for example eucalyptus and beech branches, is ideal for achieving the basic framework for a traditional pedestal display. It can be used to fill the gaps between round and linear materials, and adds grace and depth to the design.

This traditional pedestal display combines every color, form, and texture to stunning effect. Cranberry branches and spray foliage are used to link the round and linear materials, blending the two together and softening the overall appearance. Large bunches of grapes, persimmons, and loquats are centered towards the middle of the display to create a focal point, drawing the eye in.

TEXTURE

It is not only color and form, but also the texture of plant materials that plays a predominant part in how a display is perceived. With artificial, preserved, or dried materials in particular, the choice of textures is endless: for example, the rough, matt and woody feel of lotus heads; the porous, spongy appearance of mushrooms; and the glossy, smooth look of the apple. A combination of these contrasting textures can only serve to enhance any design.

In any good design, visual textures should be as varied as possible, particularly when creating a monochromatic design. The juxtaposition of shiny and matt, and velvety and smooth textures, can be carried through with rewarding results in displays of all kinds. In a dark corner of a room, you can create the illusion that more light is present by choosing the correct texture combination.

Once you have mastered the art of selecting the correct color, form, and texture for your chosen design, the actual arranging and assembly of the display is remarkably easy to achieve.

This posy combines round materials with a circular design to produce a pleasing sense of symmetry. The different-size fruits and berries add volume and help to draw the eye into the center of the circle.

The use of a linear material such as chilies can add animation to an arrangement. Here, they are positioned so that they point in all directions, creating a strong sense of passion and vitality.

This monochromatic display contains primarily neutral, earthy colors. The impact of the arrangement comes from the variety of visual textures: the rough, cracked bark contrasts with the smooth, streaky lines of the driftwood and the glossy gourds stand out against the matt surface of the sponge mushrooms.

Getting Started

It is important to grasp the basic principles of design before attempting any of the projects that follow in this book. The first few chapters will provide you with the foundation knowledge to begin arranging with confidence.

Whatever your level of skill in arranging materials, the single and most important rule to follow is that you should always choose materials which you personally find aesthetically pleasing. You should then try to arrange them in a way that will enhance their natural grace and beauty. Do not try to be too clever. Simplicity is the key and over-arranging is a common pitfall, making the end result look rigid and inanimate. For your first attempt at arranging, limit the number of colors and variety of materials you use.

Never rush into a design. There are certain factors to consider before embarking on any project, namely: (a) the environment in which the arrangement is to be placed, (b) the background against which it is to be set, (c) the color scheme, and (d) your own budget. If you start arranging without considering these things, the result will be confused and unsatisfactory.

Make sure you have mastered the basic techniques. Even when you have completed several projects successfully, continue to refer to the techniques section to avoid any disappointment (even if only to refresh your memory and, therefore, to increase your confidence). Choose a work area with good lighting and a reasonable sized work bench – possibly a kitchen table or a garage work bench. Bear in mind that once

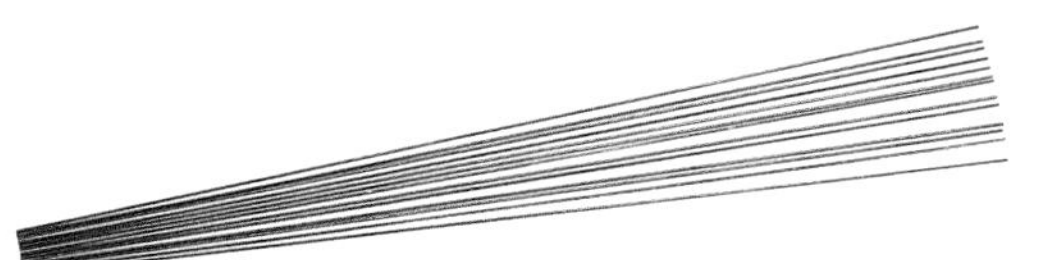

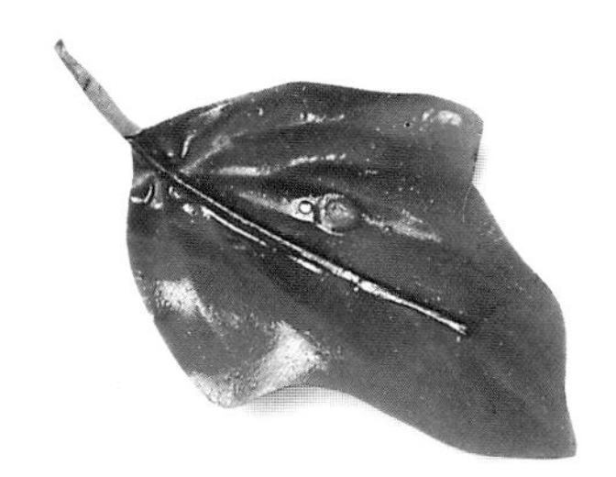

you are in the throes of an arrangement, the mess factor can become out of control. Try to discipline yourself by tidying up as you go along, and, wherever possible, keep trimmings for future use.

The projects throughout this book can be followed to the last detail or adapted to suit your own personal needs and tastes. Where large quantities of fruits and vegetables are required or where the quantity affects the design of a display, we have indicated how many of each item you will need. However, in many cases it will be a matter of choosing a selection of fruits and vegetables from those available to you; bunches of grapes come in different sizes as do most fruits and vegetables, so you must use your own judgement and regard our recommendations as guidelines only. Do not be afraid to veer off the track a little and substitute one material for another or use an exciting or unusual container. Use your imagination!

ARRANGING TIPS

Always have a good selection of differing thicknesses of wire. You will have to use your own common sense in deciding which thickness to use for each item. Bear in mind that heavy, large-headed fruits and vegetables require a thicker gauge of wire, whereas you could get away with using a medium-to-thin gauge for smaller-headed fruits, such as strawberries, cherries, and also for leaves and foliage. As a general rule, you will find that the thinner the wire, the more flexible the stem.

Often when you first buy your artificial materials, you will find that they have been so tightly packaged that the leaves on the stems have become crushed and distorted. A little gentle manipulation will soon restore them to their former glory. When opening the leaves out, check that they are not all pointing in the same direction. Follow the rules laid down by nature and alternate the direction in which the leaves should be facing.

Everything can be used to good effect – even the smallest of leaves is useful – so never discard any cuttings. If you do not use them immediately, store them carefully so that they can be put to good use later in a different display. The same rule applies to floral foam. Never throw away any offcuts as these can be used as wedges to secure a larger piece of foam more securely into a container.

Use foliage to give a basic framework and outline in large arrangements, and as a filler in smaller displays.

Always cut artificial stems with a good wire cutter or pair of pliers. Invariably, the stems will have a length of wire running through the middle, which is difficult to cut with a normal pair of scissors. The scissor blades will become damaged if used in this way.

Dressing for Dinner

Special dinner parties call for the devising of coordinated decorations to put the finishing touches to good food and good company. The look you aim for will depend on the time you have available and the type of gathering you are planning. Many of these settings can be planned and prepared in advance, leaving you free to concentrate on your guests – and on the important business of enjoying yourself!

Floating Islands

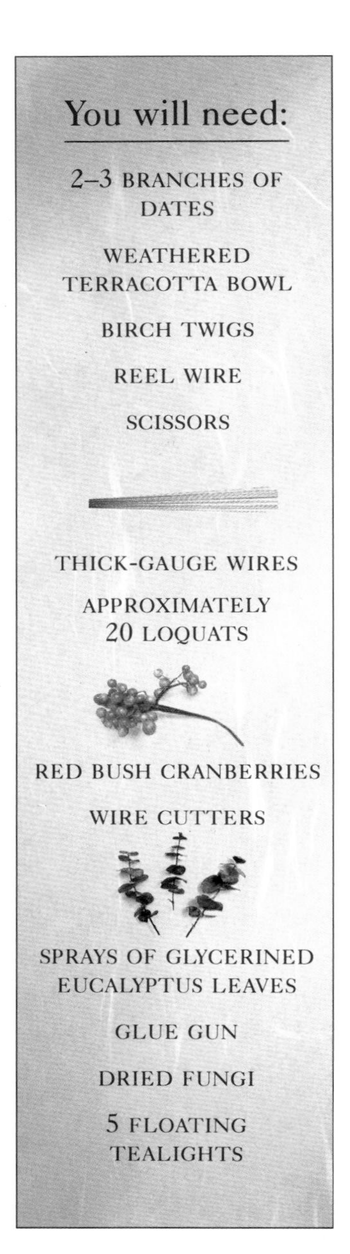

You will need:

2–3 BRANCHES OF DATES

WEATHERED TERRACOTTA BOWL

BIRCH TWIGS

REEL WIRE

SCISSORS

THICK-GAUGE WIRES

APPROXIMATELY 20 LOQUATS

RED BUSH CRANBERRIES

WIRE CUTTERS

SPRAYS OF GLYCERINED EUCALYPTUS LEAVES

GLUE GUN

DRIED FUNGI

5 FLOATING TEALIGHTS

A simple terracotta bowl can be transformed into a magical centerpiece for a special dinner party. The deep, rich colors of the garland make a perfect foil to the muted shades of terracotta and the glowing tealights add the finishing touch.

1 Shape the date branches into curves and twist them around one another to form a circular garland, which will sit comfortably around the circumference of the terracotta bowl.

2 Take a small bunch of birch twigs and shape them into a curve. Twist the birch twigs around the date garland framework.

3 With reel wire, secure the birch twigs into position. Follow the curve of the birch twigs, taking care to ensure that each end is strongly secured. Trim off any excess birch twigs with scissors.

4 Wire up the fruits and berries. The loquat fruits should be positioned either singly or in pairs, while the cranberries should be grouped in large clusters.

5 Attach the fruits and berries evenly on to the garland by twisting the wire stems into and around the date branches and birch twigs. Carefully place each fruit and berry to maintain the arrangement's symmetry. Use wire cutters to trim off any excess wire.

6 With scissors, cut off several small sprigs of eucalyptus leaves. Start filling in any remaining gaps between the fruits and berries, securing the leaves with glue. Follow the same method to glue the dried fungi onto the garland.

7 Position the garland over the rim of the terracotta bowl. Fill the bowl with water. Place floating tealights on the surface of the water and light them just before your guests come to the table.

Natural Landscape

You will need:

2 FLORAL FOAM BLOCKS

BLACK PLASTIC TRAY

FLORISTS' TAPE

SEVERAL PIECES OF DRIFTWOOD

3 LOG TEALIGHTS

WIRE PINS

THICK- AND MEDIUM-GAUGE WIRES

BARK OF VARYING LENGTHS

SPONGE FUNGI

SHARP KNIFE

GREEN ONIONS

CORNCOBS

CARDAMON PODS

BULBS

BUN MOSS

GOURDS

Natural materials, such as driftwood, fungi, tree bark, and mosses, form the basis of this unusual, earthy landscape display. Colors are kept to a minimum, since it is the play on form and texture that creates the subtle, yet dramatic effect. The pretty tealights and varied heights of grouped materials can be reflected to spectacular effect if you place them in front of a mirror. For best results, create this arrangement in situ as not all the elements are attached securely.

1 Place two blocks of floral foam on top of one another in the black plastic tray. Secure the foam in position with florists' tape.

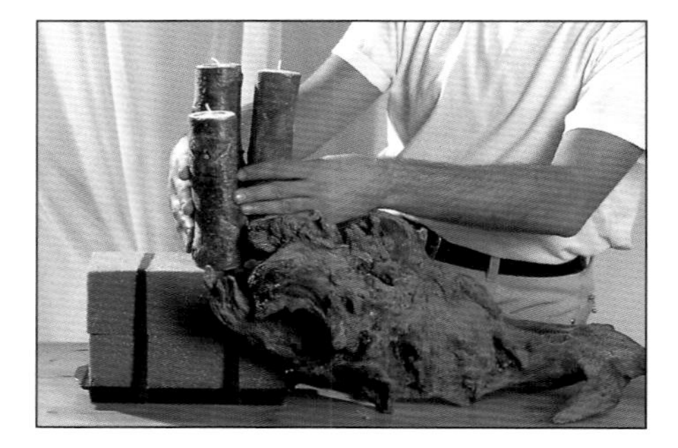

2 Position the largest piece of driftwood over one half of the floral foam. Determine the height and positioning for the log tealights by placing them on the arrangement.

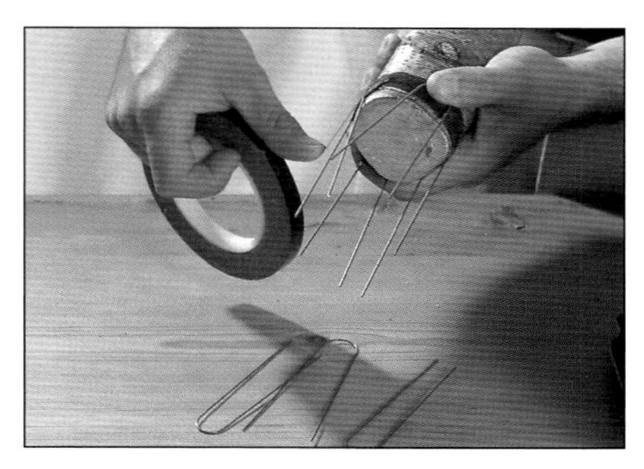

3 Secure the log tealights into position by taping several wire pins around the base of the log to form pegs. Push the pegs firmly into the floral foam.

4 With thick wire, wire up several pieces of bark of varying lengths. Insert the bark into the floral foam at the opposite side to the large driftwood piece to create a visual balance in the arrangement.

5 Place smaller pieces of driftwood at the front of the display, taking into account their shape and size. With medium wire, wire up several pieces of sponge fungi of varying sizes. Place the fungi in appropriate positions around the log tealights and driftwood.

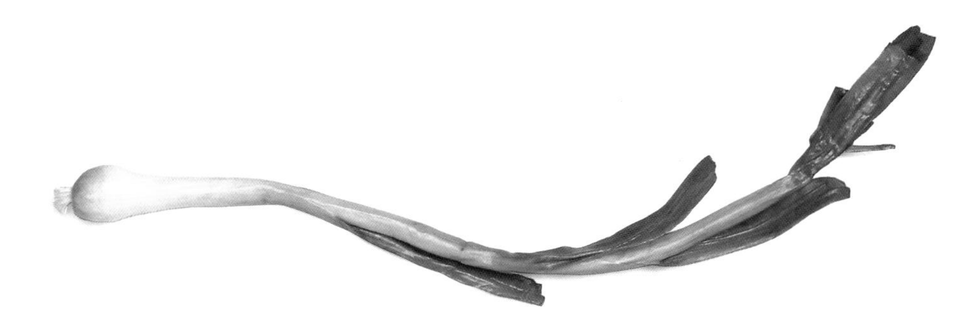

6 With a sharp knife, cut the green onions in half and position the heads at the front of the display. Twist the green onion tails at an angle to give the impression that they are growing through the arrangement. Insert the corncobs between the bark pieces to give a grouped effect.

7 Fill in any obvious gaps in the arrangement with small groupings of pods and bulbs. Where necessary, conceal any visible wiring with small clumps of bun moss.

8 As a finishing touch, place a grouping of gourds to one side of the display at the front. This will balance the corn and green onions at the opposite side of the arrangement.

Market Garden

You will need:

2 FLORAL FOAM BLOCKS
RUSTIC-STYLE BASKET WITH HANDLE
WIRE PINS
THICK- AND MEDIUM-GAUGE WIRES
WIRE CUTTERS

GREEN AND RED CABBAGES
POTATOES
CHILIES
CARROTS
BELL PEPPERS
EGGPLANTS
CUCUMBERS
ZUCCHINI
LETTUCE
GREEN ONIONS
GARDEN TWINE

As fresh garden flowers are not abundant in late fall and winter, fruits and vegetables come into their own, providing wonderful opportunities for creative floristry. This ornate rustic basket makes the most of the contrasting colors, shapes, and textures of different vegetables. Let the shapes and shades of your materials inspire you to create a display that looks good enough to eat!

1 Place two blocks of floral foam inside the basket and wire them together with wire pins. Secure by passing a length of wire through the basket and stretching it diagonally across the width. Press the wire into the floral foam, pulling it tightly, then twist the wire ends around themselves and trim off any excess. Repeat with the opposite diagonal.

2 Wire up the green and red cabbages first as they are the largest vegetables. Place the cabbages at the center of the arrangement to form the focal point. Open out the leaves to create your desired shape.

3 Continue to wire up the other varieties of vegetables. Decide carefully where each one should be placed for best effect, taking into account their color, form, and various textures.

4 To fill in any small gaps in the arrangement, wire up individual lettuce leaves. Insert them between the vegetables, covering up any unsightly floral foam.

5 As a finishing touch, attach a bunch of green onions to the basket handle with twine. Shape the leaves with your hands to create an eye-catching talking point.

Compliments of the Season

You will need:

FLORAL FOAM BLOCK

BLACK PLASTIC TRAY

SHARP KNIFE

FLORISTS' TAPE

3 CHURCH CANDLES

THICK- AND MEDIUM-GAUGE WIRES

3 SMALL TERRACOTTA POTS

SMALL PIECE OF WOOD

GLUE GUN

THIN PIECES OF BARK

SPRAYS OF GLYCERINED BEECH AND EUCALYPTUS LEAVES

BUNCHES OF GRAPES

PLUMS

RASPBERRIES

ILEX BRANCHES

CARPET MOSS

WIRE PINS

6–7 DRIED ROSE STEMS

Low, front-facing displays are perfectly suited to side tables and buffets. This woodland-inspired creation combines differing textures with the subtle colors and fruits of the fall for a charming seasonal spectacle.

1 Place a block of floral foam into the tray. Trim the edges to form a gently domed shape. Secure the floral foam in position with florists' tape at the top and bottom.

2 Insert the candles at varying heights towards the back of the floral foam. Wire up the terracotta pots and place one in front and two on either side of the candles. Glue a small piece of floral foam inside each of the pots.

3 Start building up the basic shape using long, thin pieces of bark and sprays of beech and eucalyptus leaves. Make sure that you cover the whole of the foam evenly.

4 Using the fruit as a focal point, place a large bunch of grapes between two of the terracotta pots on one side. Group plums and raspberries between the two pots on the other side.

5 Introduce some ilex branches at the back and on one side of the arrangement to add interest and more color. This will help to balance the raspberries on the opposite side as well. Cover the base of the pots with carpet moss, securing it into place with wire pins.

6 Fill each pot with dried roses. Arrange the roses to look as if they are growing naturally.

Table Garland

Garlands are an ancient form of decoration and a wonderful way of dressing large areas of space and awkward places. You can wind them around banisters, spiral them up a pillar, or drape them around doorways or tables. Wherever you use them, their effect is always dramatic and extravagant.

1 Carefully pull out all the vine leaves and grape bunches to accentuate the thickness of the trails. Twist the two vine trails together to increase the density of the garland.

2 To add interest and body to the garland, introduce small clumps of birch twigs wired into bunches. Attach the twigs at intervals between the grape bunches.

3 Prepare your fruits and vegetables by wiring each one separately with medium wire. Wire each item onto the main garland at regular intervals, alternating fruits with vegetables so that the result is symmetrical.

4 Continue adding fruits and vegetables until the length of the garland has been evenly covered. Tidy up the mechanics by trimming off any excess wires close to the binding point. Twist the garland into a circle and place it around the table centerpiece.

Terracotta Treats

You will need:

- SHARP KNIFE
- FLORAL FOAM BLOCK
- 2 WEATHERED TERRACOTTA POTS
- GLUE GUN
- BEESWAX CANDLE
- CARPET MOSS
- WIRE PINS
- MEDIUM- AND THIN-GAUGE WIRES
- SCISSORS
- DRIED FUNGI
- DRIED POMEGRANATES
- CRANBERRIES
- NUTS
- DRIED RED CHILIES
- SPRIGS OF GLYCERINED EUCALYPTUS LEAVES

Impress your guests with original ideas for place settings, using weathered terracotta as a starting point. The rich colors of the clustered fruits in one of the place settings are a combination of cranberries, pomegranates, and chilies, studded with fungi – a striking counterpoint to the gentle, tapering shape of the other display.

1 With a sharp knife, cut and shape the floral foam into two pieces. Make sure that these fit snugly into the terracotta pots, and then glue them into position.

2 Glue the end of the beeswax candle, and insert it securely into the center of one of the pots.

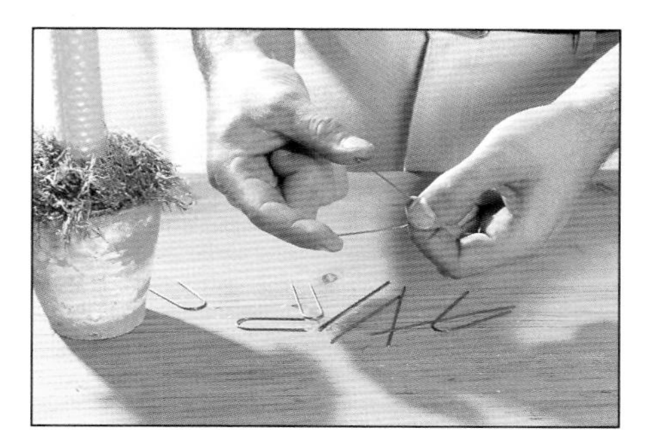

3 Surround the candle with a layer of carpet moss and secure with wire pins. Trim away any excess moss with scissors.

4 Wire up the remaining materials, then insert a grouping of fungi along the rim of the second pot. Allow the fungi to bend over the edge of the pot to look as if they are growing naturally from that point.

5 Build up a dome shape, adding one material at a time. Place the pomegranates and cranberries slightly off-center at the top and side of the arrangement to form the focal point.

6 Position the nuts at the opposite side of the arrangement to the pomegranates. This creates a balance and preserves the rounded, domed shape of the display.

7 Insert the dried chilies at the opposite side to the red berries. Fill in any remaining noticeable gaps with small sprigs of eucalyptus.

Special Occasions

At certain times of the year, you will want to create something special, a display that will make a lasting impression. Annual celebrations, particularly Christmas, provide us with the perfect excuse for decorating our homes and surroundings with stunning and extravagant displays.

Christmas Tree

You will need:

SILVER GALVANIZED BUCKET

3 FLORAL FOAM BLOCKS

SHARP KNIFE

WIRE PINS

CARPET MOSS

GLUE GUN

BIRCH TWIGS

REEL WIRE

ORANGE, LEMON, APPLE, AND PEAR FRUIT SLICES

THIN GOLD STRING

WIRE-EDGED GOLD RIBBON

MEDIUM-GAUGE WIRES

Elegant and eye-catching, this magical tree is simplicity itself. Birch twigs are decorated with slices of citrus fruits, apples, and pears. And to finish, this tree is topped with a beautiful bow and placed in a glittering, galvanized bucket.

1 Pack the bucket tightly with approximately three blocks of floral foam. Trim away any excess foam with a sharp knife. To stabilize the blocks, wire them together with wire pins.

2 Cover the top of the bucket with carpet moss. Secure this in position with wire pins and glue. Insert birch twigs evenly into the carpet moss layer, checking that the twigs are held securely by the foam.

3 Bunch the tops of the twigs together by gathering them into your hand. Wire them tightly together with reel wire, leaving a tuft of twigs at the top.

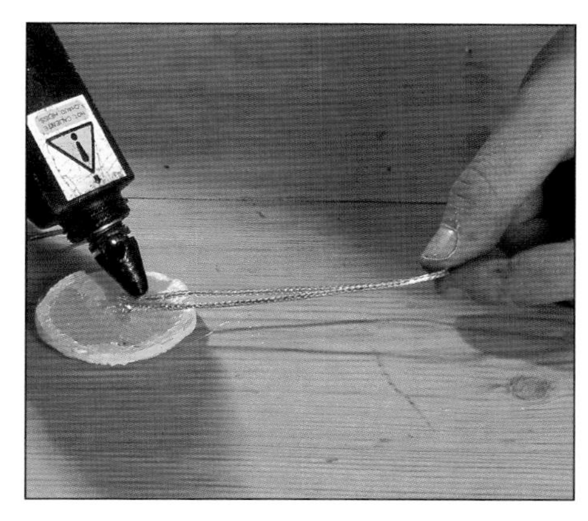

4 Prepare the fruit slices by gluing a loop of gold string onto each individual slice. Allow to dry for about five minutes.

5 Hook each fruit slice over the protruding pieces of birch twig. Check that the outside and middle of the arrangement are evenly covered to create depth.

6 To finish, make a lavish gold bow and attach this with medium wire to the top of the tree to cover the binding point.

Christmas Wreath

This rich and varied wreath features many surprise ingredients. In addition to figs, cones, and pears, the blue pine surround is studded with groups of natural beeswax candles, cinnamon sticks, and oranges.

You will need:

COPPER WIRE WREATH FRAME

BALL OF STRING

SPAGHNUM MOSS

SCISSORS

2–3 BLUE PINE BRANCHES

REEL WIRE

MEDIUM-GAUGE WIRES

3 ORANGES

3 PEARS

3 FIGS

PINE CONES

CINNAMON STICKS

BEESWAX CANDLES

GARDEN TWINE

LOTUS HEADS

SPONGE FUNGI

BIRCH TWIGS

1 Take the copper wire frame and attach the ball of string to one side. Begin padding the moss onto the framework, securing it into place by twisting the string around it.

2 Continue until the circumference of the frame has been covered by a thick padding of moss. Tie off the string firmly and cut with scissors.

3 Take two to three branches of blue pine and snip off as many sprigs as you can. Attach reel wire to the copper frame and begin to build up a thick blue pine garland by staggering the sprigs around the circumference at short intervals. Secure them in place by twisting the reel wire around each pine sprig.

4 Continue to build up the blue pine garland until the circumference has been covered. The end result should be an even and abundant ring of foliage.

5 Wire up the fruits with medium wire. Space them evenly around the ring, three oranges at the base, three pears to one side, and three figs to balance the arrangement on the opposite side. Each individually-wired piece of fruit can easily be pushed through the frame. Twist the wire stems around one another to secure the fruits in position.

6 Place small groups of wired pine cones between the fruit, then begin to make parcels of cinammon sticks and candles. These are tied together with garden twine and wired onto the main framework.

7 To add interest and another texture, insert large lotus heads and sponge fungi in the gaps. Finish off by inserting twisted branches of birch twigs into the arrangement as an unusual substitute for ribbon.

8 To hang the garland on a door or wall, make a wire hook with reel wire and attach this to the original copper frame with more wire.

Festive Centerpiece

You will need:

- WET FLORAL FOAM BLOCK
- BLACK PLASTIC TRAY
- FLORISTS' TAPE
- THICK BEESWAX CANDLE
- HEAVY-GAUGE WIRES
- BLUE PINE SPRIGS

- NATURAL PINE CONES
- VARIEGATED HOLLY WITH BERRIES
- 3 RED APPLES
- 3 RED PEARS

- 2 GOLD-SPRAYED GRAPE BUNCHES
- WIRE-EDGED GOLD RIBBON
- ILEX SPRIGS

The Christmas meal is incomplete without a stunning centerpiece to enhance the festive atmosphere. Shimmering golds, combined with deep red apples and cool pine-scented greens, are a winning choice for this season of goodwill.

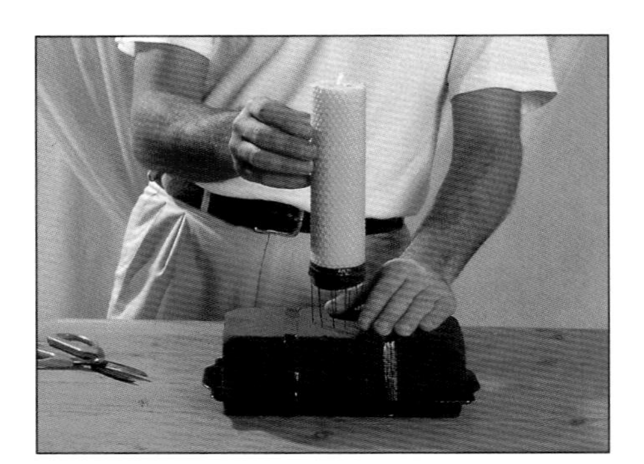

1 Place a block of wet floral foam inside a shallow, plastic tray and secure it to the tray with florists' tape. Next, wire up the church candle and wedge this into the foam.

2 Insert varying lengths of blue pine sprigs around the base of the display to achieve the desired shape. Check that the foliage radiates evenly around the foam base. Add shorter pieces of blue pine to the top and sides of the arrangement until the foam is completely covered.

3 Place sprigs of holly in between the blue pine pieces and evenly throughout the display. This forms the basis of the arrangement.

4 Wire up your apples, pears, and grape bunches. Position the three apples at the base of the candle on one narrow side of the display and the three pears on the other narrow side. Use the grapes to give a tapering effect on the two long sides of the display. Wire up your natural cones and use these to fill in any obvious gaps between the apples, pears, and grapes.

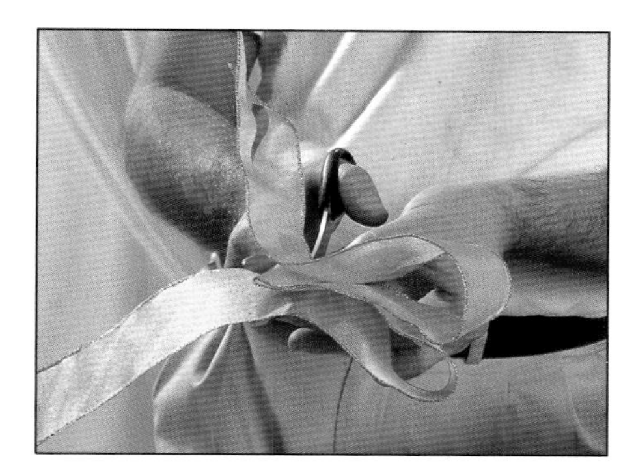

5 Prepare four bows and two ribbon trails from the gold ribbon. Wire two bows in place on either narrow side of the display, with the two ribbon trails at the longer ends to accentuate the tapered line of the display.

6 To finish, place small sprigs of ilex evenly throughout the display. These are totally flexible, and can be molded and curved to the desired shape and direction.

Christmas Garland

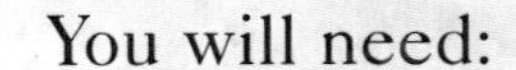

You will need:

ROPE

SCISSORS

REEL WIRE

BLUE PINE SPRIGS

ILEX BRANCHES

THICK- AND MEDIUM-GAUGE WIRES

DRIED POMEGRANATES

GOLD-SPRAYED GRAPE BUNCHES

GOLD-SPRAYED LOTUS HEADS AND PINE CONES

WIRE-EDGED BURGUNDY RIBBON

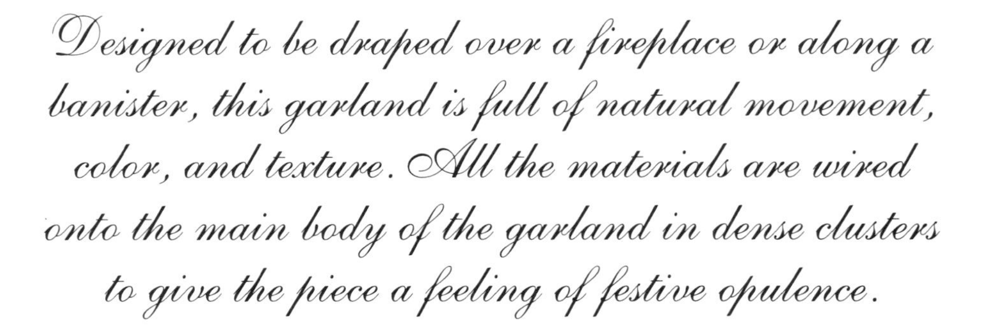

Designed to be draped over a fireplace or along a banister, this garland is full of natural movement, color, and texture. All the materials are wired onto the main body of the garland in dense clusters to give the piece a feeling of festive opulence.

1 Decide how long you wish the garland to be and cut a length of rope accordingly. Make a small loop at either end to allow you to attach it to your chosen area later on. Attach the reel wire just below one of the loops. Prepare your sprigs of blue pine – those with three prongs (shaped like a fork) are ideal for this purpose.

2 Attach the first sprig of pine onto the string and bind the reel wire twice around the base. Check that the string loop is adequately concealed. Continue to stagger pine sprigs along the length of the twine, alternating the sides and rotating the body of the garland as you go along. Each piece must be wired individually into position with the reel wire. Once you have reached the end of the rope, break off the wire.

3 Ilex branches are very flexible and can be pushed into the body of the garland, then twisted around the length of it to secure. The red berries should be spaced evenly across the garland.

4 Wire up the pomegranates and grape bunches. Place them in clusters of three at regular intervals along the length of the garland. The wire stems should be pushed through the body of the garland, then bent upwards and back around the garland to secure.

5 Next, wire up the lotus heads and pine cones and attach them in groups of three at regular intervals along the garland.

6 Finish off by placing bows between the materials, staggering them as you go. Finally, pass a loop of ribbon through the twine loops at either end to make a larger, more attractive loop to hang the garland.

Christmas Vase Display

With the most basic of ingredients and a little imagination, you can create a stunning vase display to enliven a hallway or living room. Its simplicity and informality serves to enhance the natural beauty of the seasonal foliages used.

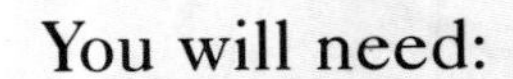

You will need:

GREEN GALVANIZED BUCKET

BLUE PINE BRANCHES

HOLLY BRANCHES

SPRAYS OF GLYCERINED RED OAK LEAVES

RED AND YELLOW DOGWOOD TWIGS

ILEX BRANCHES

1 Fill the bucket three-quarters full with water. Take the blue pine branches first as these are the heaviest, bulkiest foliage, and begin to build up your required shape. Keep in mind the position of the vase once the arrangement is complete and the fact that any vase display should stand one and a half times the height of the container.

2 Once the basic width, height, and density has been achieved with the blue pine, fill the gaps with holly branches. Use the longest pieces to add height and the shortest pieces to fill out the front and middle of the arrangement. Note the natural curvature of each branch and place it accordingly.

3 To add color and another texture to the display, insert a few branches of red oak leaves evenly throughout the arrangement. These branches are quite expensive, so use them sparingly. To give the display an even more festive feel, introduce some twigs of red and yellow dogwood.

4 To finish, place branches of ilex liberally throughout the display. This not only gives the arrangement a Christmas feel, but also adds color and texture.

Heart-to-Heart

This delightful and unashamedly romantic interlocking double-heart basket makes an original gift for Valentine's Day, or a special anniversary. One basket is packed with head-to-head dried rose blooms; the other is piled high with strawberries and raspberries.

You will need:

CHICKEN WIRE

WIRE CUTTERS

CARPET MOSS

GLUE GUN

SCISSORS

FLORAL FOAM BLOCK

SHARP KNIFE

LARGE BUNCH OF HEAD-TO-HEAD DRIED RED ROSE BUDS

STRAWBERRIES

RASPBERRIES

RED BUSH CRANBERRIES

1 Take a length of chicken wire about 12 in x 24 in (30 cm x 60 cm). Halfway along the length, snip about 3 in (7.5 cm) into the wire at both top and bottom.

2 Mold the wire with your hands into two interlocking heart shapes by bending the mesh over to form a point at the bottom and two curves at the top.

3 Take a handful of moss and squeeze a generous helping of glue onto the underside. Press the moss carefully onto the framework.

4 Continue pressing a handful of moss at a time, maintaining the heart shape as you go, until the wire frame is completely covered.

5 Trim off any excess moss with scissors to give a more defined outline to the heart shape.

6 Cut a heart shape out of floral foam and place it in one side of the basket. Fix the heart-shaped foam into position with glue.

7 Cover the floral foam with dried rose buds. Use the smallest buds for the outer edges and the larger ones towards the center of the heart. In this way you create a domed, cushioned effect.

8 Loosely fill the second section of the heart basket with a generous helping of strawberries, raspberries, and red bush cranberries. Allow the berries to spill over the edge slightly for a more natural look.

Hot Passion

You will need:

SCISSORS

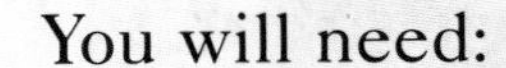

SPRAYS OF GLYCERINED EUCALYPTUS LEAVES

GLUE GUN

TWIG WREATH FRAME

REINDEER MOSS

DRIED CHILI PEPPERS

Bright, shiny chilies are suitably striking tokens of love in this eye-catching wreath. Interlaced with eucalyptus and reindeer moss, this arrangement is a fitting symbol for red hot lovers everywhere!

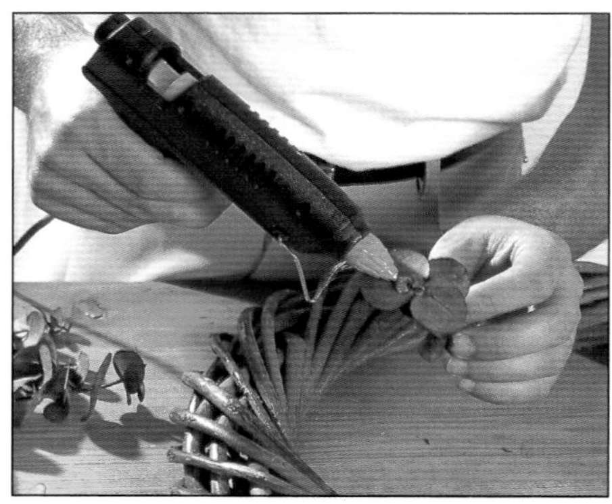

1 With scissors, snip off a small sprig of eucalyptus. Squeeze a small amount of glue onto the end of the sprig. Press the sprig lightly onto the twig framework of the wreath, positioning it at an angle.

2 Take a small clump of reindeer moss and squeeze a small amount of glue onto the underside. Press the moss lightly onto the twig wreath, close to the eucalyptus sprig.

3 Glue a chili pepper between the moss and the eucalyptus sprig in the same way. Take into account the natural curve of the chili, which should be pointing outwards.

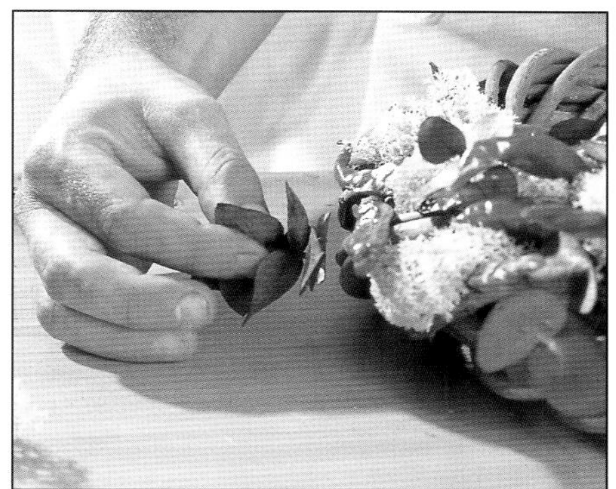

4 Work your way around the wreath, alternating the three materials as you go, until it is fully covered. The overall effect of this symmetrical arrangement is both decorative and eye-catching.

Halloween Centerpiece

You will need:

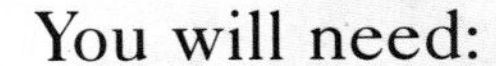

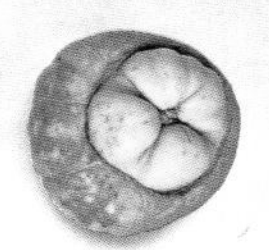

3 LARGE GOURDS

BRANCHES OF LICHEN TWIGS AND DRIED OAK LEAVES

REEL WIRE

CARPET MOSS

Delve deep into the forest for the inspiration to make this stunning arrangement for fall. The theme is Halloween, so gourds make a natural centerpiece. Carefully placed within a nest of twigs and crisp oak leaves, they create an autumnal glory, framed beautifully by a cage of lichen-tinged larch in a glorious medley of rich, golden colors.

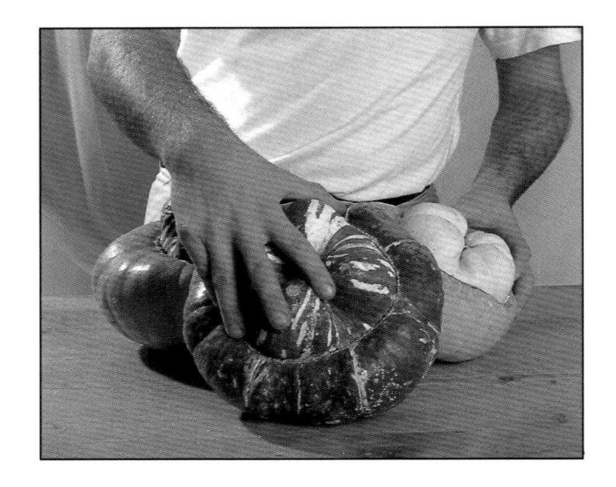

1 Select three gourds of different shape, color, and size. Place them together carefully, one balanced against the other.

2 Make a garland to surround the gourds out of lichen twigs and branches of dried oak leaves. Wire the two together, shaping and sizing the garland so that it fits snugly around the central cluster of gourds.

3 Insert small clumps of moss between the twigs and gourds to give a more natural, "forest" feel to the arrangement and to disguise any wires.

4 To complete the eerie effect, create a light cage of thin lichen twigs rising up and bending gently over the arrangement. Insert the ends of the twigs loosely into the garland.

Harvest Wreath

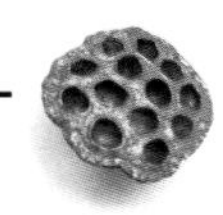

You will need:

3 BEESWAX CANDLES

FLORAL FOAM WREATH

WIRES OF ALL GAUGES

4–5 SPONGE FUNGI

GLUE GUN

REINDEER MOSS

CORNCOBS

SUNFLOWERS

CARTHAMUS

GOURDS

LOTUS HEADS

BUN MOSS

WIRE PINS

WHEAT

SPRAYS OF GLYCERINED EUCALYPTUS LEAVES

This abundant table wreath brings together the harvest fruits of nature. Corncobs, sponge fungi, gourds, wheat, and lotus heads are all carefully grouped for maximum effect – the perfect arrangement to grace any festive table.

1 Take three candles of varying heights and push each one into the floral foam wreath. Position them close together, so that they are touching one another.

2 Wire up four or five pieces of sponge fungi and place them around the bottom of the candles so that they look as if they are growing naturally from this point. Glue small pieces of reindeer moss over the wires to conceal them.

3 Group three corncobs directly opposite the candles to create a balance of color and form. Start to build up your groupings of other materials. Decide where each item should be placed before inserting it into the arrangement.

4 Fill in the gap between the gourds and the lotus heads with bun moss, secured into position with wire pins. Cut the wheat quite short and use this to fill the gap between the candles and the gourds.

5 To finish, cut small sprigs of eucalyptus and push them into the floral foam, wherever there is an obvious space, or where the wires need to be concealed.

Topiaries and Sculptures

Topiaries and other plant sculptures are traditionally regarded as formal arrangements, yet today's designs are more likely to be relaxed and stylishly simple. Whether the finished display is a pyramid of peaches or a sphere of nuts, you cannot avoid making an impact.

Peach Pyramid

You will need:

2 FLORAL FOAM BLOCKS

BLUE FRUIT BOWL

GLUE GUN

SHARP KNIFE

WIRE PINS

CARPET MOSS

THICK- AND MEDIUM-GAUGE WIRES

APPROXIMATELY 20 PEACHES

BUNCHES OF BLACK GRAPES

RASPBERRY CLUSTERS

The ancient Egyptians were the first to realize the eye-catching qualities of pyramids. This pyramid-shaped display is easy to create, requiring nothing more exotic than a beautiful glass fruit bowl. Berries, laced between the velvety peaches, add depth and color to the arrangement.

1 Place the blocks of floral foam centrally together inside the fruit bowl. Secure into position at the base with glue.

2 With a knife, roughly shape the foam into a pyramid. The height and width should be in proportion with the fruit bowl. Once you have carved the shape you want, wire the two blocks together with wire pins to stop any movement. Cover the area between the foam base and the rim of the bowl with moss.

3 Wire up the fruits and then begin to build up the pyramid, starting from the base. Alternately place one peach, a grape bunch, and a cluster of raspberries side by side.

4 Continue to build up the arrangement, positioning the peaches in horizontal cirlces around the pyramid. Check that the areas between the peaches are filled with symmetrical groupings of grape and raspberry clusters.

5 Check for any remaining gaps in the arrangement where the foam can still be seen and fill them with leaves from the raspberry clusters.

Classical Feast

This urn, with its sumptuous contents, has all the style and richness of a bacchanalian feast! Although the result looks complicated, it is actually quite easy to achieve, the trick being to group the fruits and berries in strict rotation, gradually building up the density of the display.

You will need:

2 FLORAL FOAM BLOCKS

CAST-IRON URN

SHARP KNIFE

WIRE PINS

CARPET MOSS

THICK- AND MEDIUM-GAUGE WIRES

BUNCHES OF BURGUNDY GRAPES

APPLES

PLUMS

KIWI FRUITS

RASPBERRIES

CHERRIES

GREEN BUSH CRANBERRIES

STRAWBERRIES

EXTRA FOLIAGE

1 Pack the blocks of floral foam tightly into the urn. Allow at least 6 in (15 cm) of foam to protrude above the rim of the urn. Trim off any excess foam with a sharp knife.

2 For added security, push wire pins through the two foam blocks at the point where they join to hold them together. Pack carpet moss tightly around the base of the urn. The moss will prevent any movement of the arrangement and also camouflages any unsightly remaining foam.

3 Wire up bunches of grapes and place the first bunch at the base of the urn, allowing it to tumble over the edge. Push the wire stem deep into the floral foam to secure it. To create a grouped effect, place the second bunch of grapes just above the first, and the third bunch just above the second.

4 Wire up the remaining fruits and decide where each grouping is to be placed. The apples will form the focal point of the arrangement and are placed at the top to form a pinnacle.

5 Begin to build up the groupings of fruit at the back and front of the arrangement so that they complement each other. Here, the two bunches of grapes at the back balance the three at the front.

6 Continue the grouping until all the spaces have been filled. Fill in any small gaps with sprigs of foliage and individual berries.

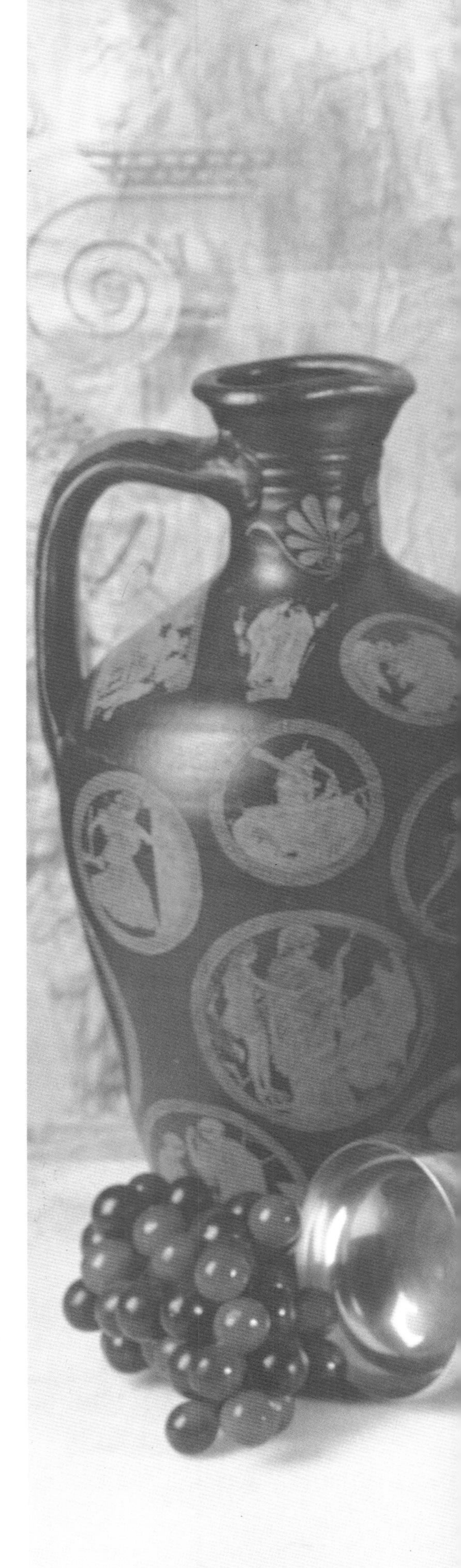

A Little Nut Tree

An ornate cast-iron pot can be transformed into an attractive little nut tree to make a really unusual, stylish decoration. Although it takes time to construct the nut tree, this impressive topiary is surprisingly easy to achieve.

You will need:

FLORAL FOAM BLOCK

CAST-IRON URN

STICK

GLUE GUN

FLORAL FOAM SPHERE

CARPET MOSS

APPROXIMATELY 150 WALNUTS

MEDIUM-GAUGE WIRES

1 Position the block of floral foam inside the container. The top of the foam should be level with the rim of the container. Place the stick into the center of the foam block, then glue it into position. The stick will act as a support for the topiary sphere.

2 Using the glue gun, cover the floral foam sphere with carpet moss. Press the sphere firmly onto the stick, leaving a gap about the size of a walnut between the sphere and the container.

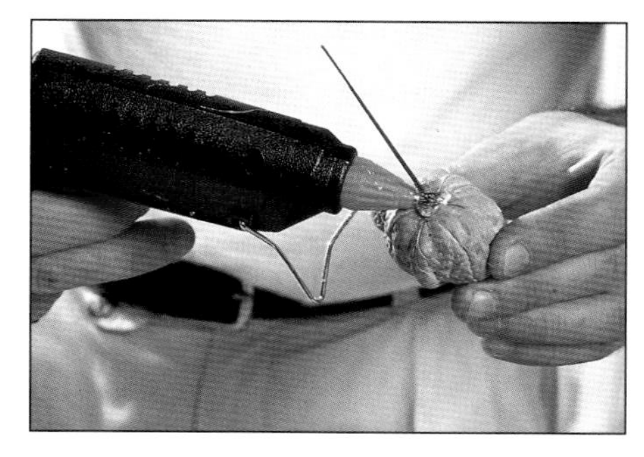

3 Wire up all of the walnuts by inserting a medium wire through the natural hole at the base of the nut. Glue into position with the glue gun and leave to dry.

4 Insert the wire stems of the walnuts into the sphere, pressing firmly until each walnut is level with the moss. Work evenly over the sphere, not forgetting the underside, until the whole sphere is covered.

*W*inter Medley

You will need:

LARGE WHITE PLASTIC BOWL

STONE URN

2–3 FLORAL FOAM BLOCKS

WIRE PINS

CHICKEN WIRE

BIRCH TWIGS

GLYCERINED EUCALYPTUS AND BEECH BRANCHES

CRANBERRY BRANCHES

WIRE CUTTERS

GRAPES

PERSIMMONS

LOQUATS

RED, YELLOW, AND GREEN CHILIES

CARROTS

EGGPLANTS

PINEAPPLES

LOTUS HEADS

STICKS

MEDIUM-GAUGE WIRES

EXTRA FOLIAGE

When the occasion calls for something grand and imposing, a pedestal arrangement fits the bill perfectly. This classic urn makes an ideal base on which to build a spectacular combination of chilies, carrots, eggplants, grapes, loquats, and a mixture of berried foliages in an explosive cocktail of color.

1 Place a large, white plastic bowl inside the stone urn. Fill the bowl with the blocks of floral foam. Secure the blocks together with wire pins, then cover with chicken wire for added stability.

2 Position the longest birch twigs, and eucalyptus and beech branches first. As a general guideline, the arrangement should be at least one and a half times the height of the container. Follow the natural curves of the branches. The most arched pieces of foliage should be used to create a trail at the front and sides of the arrangement. Shorter pieces should be used to build up the density towards the center of the display.

3 Arrange the cranberry branches evenly between the existing foliage branches, taking into account the different colors of berries. Cut some of the branches in half using wire cutters, and use the shorter pieces to build up more density towards the center of the display.

4 Mount all the vegetables and fruits on sticks to create long stems. Some sticks should be shorter than others, depending on where they are to be inserted. The longest stems should be placed around the edge of the arrangement, with the shorter ones towards the center.

5 Insert the fruits and vegetables in loose groupings around the display for maximum effect. Check that the largest-headed fruits and vegetables (grapes, persimmons, and loquats) are placed in the middle to create a focal point.

6 Finally, accentuate the outline by filling in any gaps around the edge with pieces of foliage. Check that the display has an even distribution of materials.

Citrus Tree

You will need:

FLORAL FOAM BLOCK

SHARP KNIFE

NATURAL-COLORED CONTAINER

BIRCH TWIGS

THICK-GAUGE WIRES

FLORAL FOAM SPHERE

GLUE GUN

CARPET MOSS

SPRAYS OF GLYCERINED EUCALYPTUS LEAVES

APPROXIMATELY 20 LEMONS

Adding a dash of color to this arrangement of birch twigs by introducing lemons will give your creation a refreshing twist. Here, eucalyptus, birch twigs, and moss are the basic raw materials needed to make a stunning display that looks especially welcoming in a hall. Do not feel restricted in your choice of fruits – oranges, apples, or pears work just as well as lemons.

1 Shape the block of floral foam to fit snugly inside the container, making sure that the block is level with the rim. Bunch together some birch twigs to form the trunk of the tree, and push these firmly into the foam.

2 Bind the tops of the twigs with wire to hold them together. Take the foam sphere and press it centrally onto the top of the trunk. For added security, the sphere can be glued into position.

3 Cover the top of the container with moss to conceal the foam. Prepare the eucalyptus by snipping even-sized sprigs off the main branch. Insert the sprigs into the sphere in an even, symmetrical fashion, starting with a complete circle over the top.

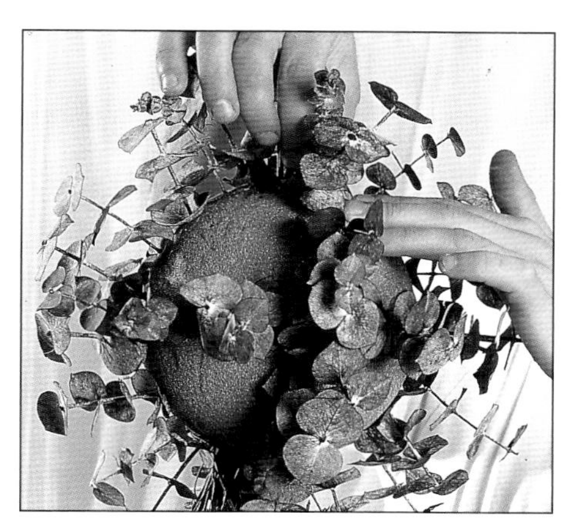

4 Continue to build up the eucalyptus base by adding a ring of sprigs around the middle of the sphere and one more over the top at right angles to the first. Finish by filling in the gaps between the segments formed by the rings.

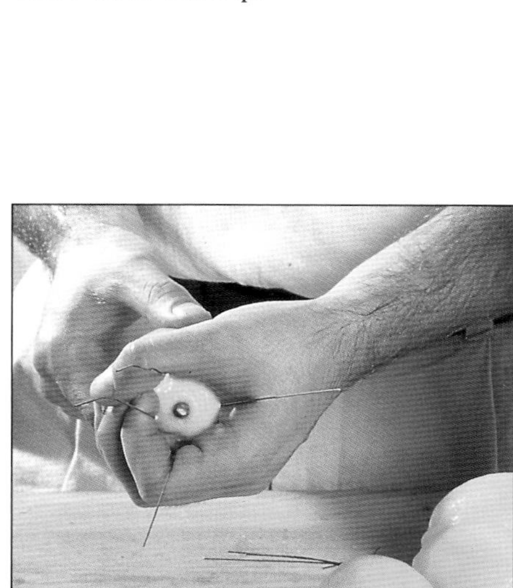

5 Wire the lemons at the base. The wire you use must be thick enough to support the weight of the fruit.

6 Insert the wire stems of the lemons deeply into the sphere at even intervals. Continue adding lemons until you have covered the tree in a symmetrical fashion.

*B*ouquets and Sheaves

A carefully constructed bouquet that embodies all the nostalgia and charm of a bygone era is always a joy to behold. A traditional sheaf has an enduring earthy power and a raw elegance that is quite enchanting. Using fruits and vegetables as the basic raw materials gives new scope for design in all shapes and styles of bouquets and sheaves, so allow your creative ideas to flow and your imagination to run free.

Victorian Inspiration

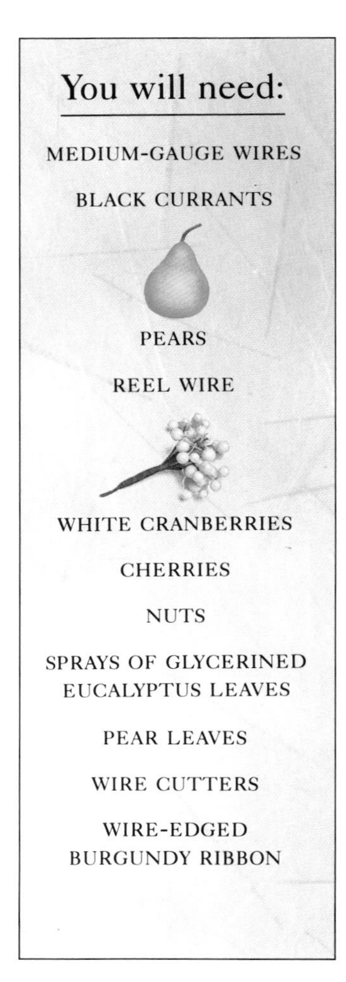

You will need:

MEDIUM-GAUGE WIRES

BLACK CURRANTS

PEARS

REEL WIRE

WHITE CRANBERRIES

CHERRIES

NUTS

SPRAYS OF GLYCERINED EUCALYPTUS LEAVES

PEAR LEAVES

WIRE CUTTERS

WIRE-EDGED BURGUNDY RIBBON

The Victorians had a fine eye for detail, so this authentic Victorian posy demands patience and time. However, the combination of fruits is both simple and original, and the perfect symmetry of the arrangement makes it timelessly classical.

1 Wire up all the materials before beginning to make the posy. Take the cluster of black currants as your focal point, and start by placing the pears evenly around its circumference until you have created a concentric circle. Position them slightly below the berries to create a gentle, domed shape.

2 Attach the end of the reel wire to the neck of the arrangement, and wrap it around the neck tightly to ensure that the materials do not move. Do not break off the reel wire.

3 To make the next circle, which consists of white cranberries, position each small berry cluster slightly beneath the heads of the pears. Bind the clusters individually into place with the reel wire.

4 Continue to bind in the remaining materials until the posy has reached the size you require. Finish off with a circlet of green leaves to frame and support the materials within the posy.

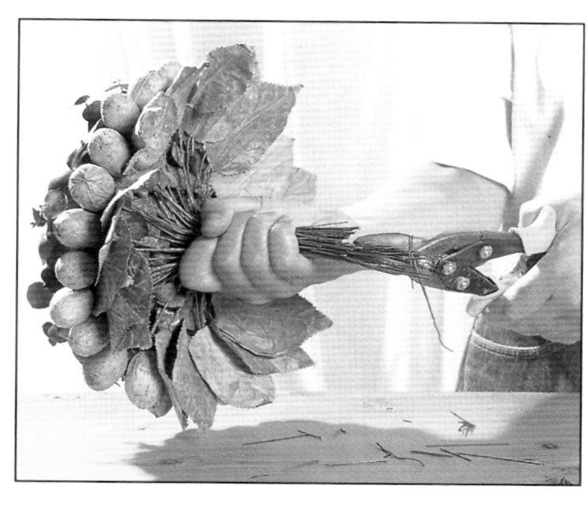

5 Break off the wire and trim the false "handle" to the required length and width with wire cutters.

6 Secure the handle with a coordinating ribbon, wrapping the ribbon tightly around the handle to conceal the wire. To finish, tie a bow at the neck of the arrangement.

Bridal Bouquet

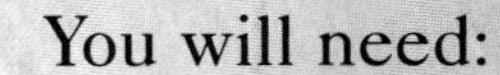

You will need:

MEDIUM-GAUGE WIRES

SPRAYS OF GLYCERINED EUCALYPTUS LEAVES

BUNCHES OF GRAPES

CRANBERRIES

CHILIES

GREEN ONIONS

PLUMS

EGGPLANTS

PINEAPPLES

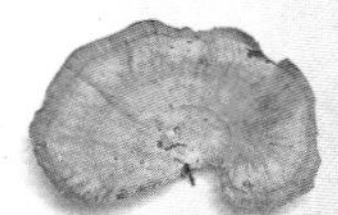

SPONGE FUNGI

GARLIC BULBS

REEL WIRE

WIRE CUTTERS

WIRE-EDGED BURGUNDY RIBBON

This unique idea for a bridal bouquet is guaranteed to tantalize happy brides. After all, how many brides would be tempted to throw a bouquet that is sure to last and last? A bold, innovative idea designed for a daring bride, the mixture of colors and textures ensures that this display will be a stunning focal point on that special day – and for many days afterwards.

1 Lay all your materials on the table and arrange them loosely into the required design. Check that the stems are the right length to form the shower effect of the design.

2 To avoid giving this traditional bridal bouquet a stiff, formal feel, wire the foliage, fruits, and vegetables from halfway down their natural stems. Those materials which are to be positioned at the top of the bouquet should be wired a quarter of the way down their stems. This will allow the spray to have more natural movement and freedom. You will have to use more than one wire with some of the fruits and vegetables to achieve the required length.

3 Take the longest stem of eucalyptus and bend the wire stem right back to form the handle of the bouquet. The point at which the wire is bent is the binding point. Tie the reel wire there and begin to build up the shower shape.

4 Add shorter stems of eucalyptus to either side of the longest piece, graduating each piece to retain the "V" shape. Bind each stem into position with the reel wire. Place a mirror in front of you so that you can keep an eye on the progress of the shape that you are building. Insert the longest stem of grapes into the bouquet to form the base point, then further branches to create a "waterfall" effect. Place a grouping of chilies to one side of the arrangement, and a grouping of green onions on the other to give balance to the bouquet.

5 Continue to add groups of fruits and vegetables to build up the arrangement. The largest, most bulbous fruits and vegetables – plums, eggplants, and pineapples – should be grouped as close to the binding point as possible to serve as a focal point.

6 Surround the underside of the bouquet nearest the binding point with large pieces of decorative sponge fungi, binding each piece into position with reel wire. These will help both to support the heavy fruits and vegetables above, and to define the general outline.

7 Cut and taper the stems to form an elegant handle. Twist the reel wire tightly up and down the false handle to secure and then break off with wire cutters.

8 Carefully wrap a ribbon around the false handle to conceal the wire, working down and then up. Fix a bow at the binding point to complete the bouquet.

Harvest Sheaf

You will need:

LARGE BUNDLE OF WHEAT STEMS

STRING

SCISSORS

SEAGRASS

The wheatsheaf is one of the cheapest and simplest of all arrangements. Twisted, domed, and tied with seagrass, this is the most distinctive of all harvest symbols. The secret of this sheaf's strength and compactness lies in the spiraling.

1 Strip each individual wheat stem of any excess foliage to give the stems a smooth, clean finish. Take a small bunch of wheat as your core and tie it tightly around the middle with string. Do not cut the string as you will need it for the next step.

2 Begin to build up the sheaf by taking small amounts of wheat and angling them across the core. Secure them by wrapping the string around the full circumference of the sheaf two or three times.

3 Angle the stems to create a spiraled, splayed effect. Remember to stagger the ears of wheat as you go to create a dome.

4 When you have achieved the required fullness, knot the string tightly and cut with scissors.

5 Finish off the wheatsheaf by wrapping seagrass around the middle several times to form a thick band. Make sure all the string is covered and then tie a large, loose bow of seagrass at the front.

Autumn Sheaf

You will need:

CORNCOBS

THICK-GAUGE WIRES

EUCALYPTUS

WHEAT

STRING

STRINGS OF GARLIC AND RED CHILIES

DRIED ORANGE ROSES

ALCHEMILLA

SUNFLOWERS

LOQUATS

SCISSORS

A HANDFUL OF RAFFIA

The inspiration for this impressive display comes from the harvest abundance that manifests itself when summer draws to a close and nature surrenders her bounty. Gather together materials that are as eclectic as the harvest itself – wheat, garlic, corn, and sunflowers – to create a rustic showpiece that is as brave a combination as it is bold. For the finishing country-style touch, this sheaf is tied with a raffia bow.

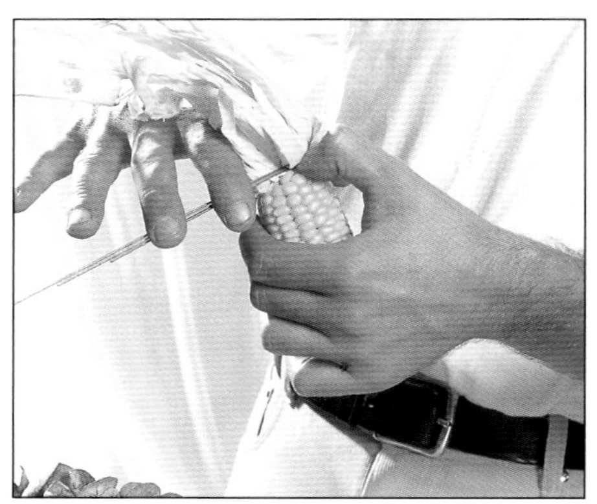

1 Prepare all the materials before starting on the hand-tied sheaf. Make a false stem for the corncobs using thick wire and clean all stems of foliage below the binding point by stripping off the leaves.

2 Take the longest pieces of foliage and wheat, and begin to cross the stems over in a spiraled manner. This creates a slightly fanned shape, and forms the base of the sheaf that will support all the other materials.

4 Insert the heaviest materials – the corncobs and the loquats – at the base of the sheaf. Bind each item in with string as you progress.

6 To complete the autumn sheaf, attach a large, loose raffia bow at the binding point.

3 Begin to build up the sheaf, inserting each material in a group for maximum effect. Bind each item in with string as you work to secure it firmly in the desired position.

5 Once all the materials have been added in, tie off the string, and trim the stems to the desired length.

Containers and Mirrors

Both containers and mirrors can be a feature of your display but remember, the size should be in direct proportion to the materials you are using and the two elements should complement one another and work together in harmony. Look around the house, garden, and surrounding countryside and you are sure to draw inspiration from the most unlikely sources! Stretch your imagination and be adventurous in your choice of materials.

Ornamental Basket

You will need:

- GLUE GUN
- BUN MOSS
- RUSTIC LOG BASKET WITH HANDLE
- DRIED FUNGI
- MEDIUM-GAUGE WIRES
- SPRAYS OF GLYCERINED OAK LEAVES
- BUNCHES OF FROSTED BURGUNDY GRAPES
- PLUMS
- BERRIED IVY
- LOTUS HEADS
- 3 STEMS OF DRIED RED AND ORANGE ROSES

Wicker baskets, especially unusually-shaped ones, are widely available and make attractive, versatile containers. Using the baskets as a framework, you can weave in mosses, fungi, berries, and fruits to enhance the basic shape and create a genuinely rustic display. Decorative as it stands, this basket also makes an ideal container for potpourri, fruits, or nuts.

1 Begin by gluing a liberal clump of bun moss to one side of the basket. Build up the arrangement around the moss by gluing a group of dried fungi in position. The fungi add an alternative color, shape, and texture to contrast with that of the moss.

2 Wire up bunches of oak leaves and attach them so that they point downwards and soften the edge. Conceal the wire by gluing on a small clump of bun moss. Wire three bunches of grapes onto the framework of the basket, following the line of the oak leaves. Conceal wires with bun moss.

3 To add more shape and texture, and continuing the color theme, attach a group of plums between the grapes and the bun moss.

4 Just off center, at the top of the basket handle, wire on two bunches of grapes to balance those on the main framework. Hide the wire with a grouping of fungi and a small clump of moss.

5 Wire up small bunches of oak leaves and berried ivy. Attach them to the base of the handle on the opposite side of the basket. They should point upwards, as though they were growing naturally. Again, conceal any wires with a small clump of bun moss.

6 Attach a small grouping of lotus heads in the same position for added interest and also to offset the heaviness of the grapes on the other side of the handle. As a finishing touch, insert three stems of dried roses into a clump of bun moss on the opposite side of the arrangement.

Banana Mania

You will need:

SHARP KNIFE

FLORAL FOAM BLOCK

YELLOW GLASS VASE

CARPET MOSS

WIRE PINS

THICK-GAUGE WIRES

APPROXIMATELY 25 BANANAS

A cascading fountain of one single fruit in a frosted yellow vase makes a colorful and eye-catching table centerpiece. Bananas have never looked more appealing than in this simple glass vase arrangement.

1 With a sharp knife, shape a piece of floral foam to fit snugly inside the vase. Cover the top of the vase completely with carpet moss to conceal the foam, securing it in position with wire pins.

2 Wire up the bananas with thick wire. Insert the first banana just inside the rim of the vase.

3 Build up the "fountain" shape by placing one banana next to the other to form a circle around the edge of the vase.

4 Make another circle of bananas inside the first, placing each fruit between the gap of the two in front. Continue to build up the circles until all the foam is covered. Finish off by placing the last banana as close to the central point as possible.

*N*ew Reflections

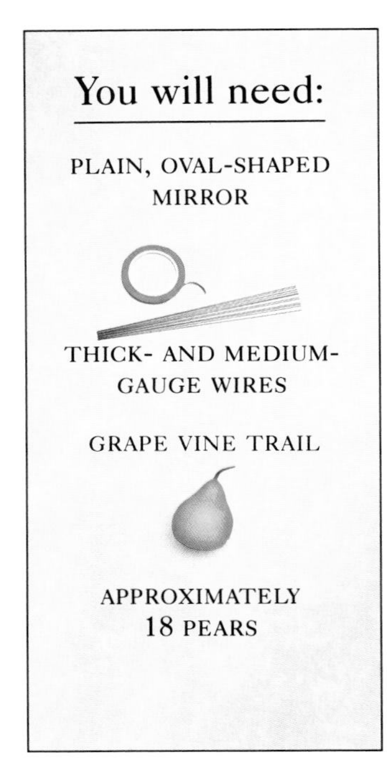

You will need:

PLAIN, OVAL-SHAPED MIRROR

THICK- AND MEDIUM-GAUGE WIRES

GRAPE VINE TRAIL

APPROXIMATELY 18 PEARS

Transform a plain, oval-shaped mirror into a sumptuous decoration with golden pears and vine fruits. The desired effect can be achieved with a minimum of materials because the denseness of the vine is enhanced by its own reflection.

1 Prepare the mirror by passing wires through the silver clips around the edge. These wires will be used to secure the vine garland in the final step. Wire up the pears in groups of two, one small pear and one larger pear.

2 Place the grapevine around the circumference of the mirror to check the correct length. Bind the two ends together with wire once you are sure of the size. Do not secure the vine to the mirror frame at this point. Instead, remove the garland and work on it separately.

3 Shape the grapes and the vine leaves so that they all point in the same direction. Begin to attach the wired-up pears to the garland at regular intervalsr.

4 Finally, position the garland back over the mirror frame and attach the two together by winding the prepared lengths of wire around the main body of the garland. Disguise any wire mechanics by bending vine leaves over the top.

Citrus Twist

You will need:

1 SMALL GLASS TANK VASE

WET FLORAL FOAM BLOCK

1 MEDIUM-SIZED GLASS TANK VASE

BUN MOSS

APPROXIMATELY 12 ORANGE SLICES

APPROXIMATELY 20 HEADS OF GERBERA

RAFFIA

FLORISTS' KNIFE

A cheat's version of topiary, this simplified "tree" uses fresh gerbera to create an illusion of exaggerated height and fullness. The orange fruit slices, encased in a moss-filled glass tank vase, provide an intriguing counterpoint to the base of the display.

1 Fill the smaller vase with wet floral foam and place it inside the larger one. Start filling the gap between the two vases with a layer of bun moss, followed by a layer of orange slices. Fill the areas between slices with moss. If the slices do not fall in a regular pattern, so much the better, as the idea is to achieve a simple and natural look. Once all four sides of the larger glass tank have been filled, begin to build up the topiary of gerbera.

2 Hold the stems as close as possible to the necks of the flowers and build up a ball shape by packing the flower heads close together. Allow the flower heads to overlap.

3 Ensure that all the stems are straight and then tie a piece of raffia around the neck of the bunch, very close to the flower heads.

4 The stems will form the trunk of the tree. Hold them tightly together and estimate how tall you would like the tree to be. Bear in mind that the height of the tree should be in proportion to the height of the container. A good guide is to make the tree twice as high as the container. Use a florists' knife to trim off the stems to an equal length.

5 Using both hands, push the stems into the wet floral foam carefully and slowly until you can feel that they are secure.

6 To finish, cover the surface area of the vase with bun moss to conceal any unsightly floral foam. To extend the life of the arrangement, water occasionally through the moss.

Mirror, Mirror

Framed by a garland of vegetables, a mirror can be a novel and unusual conversation piece! The secret of this arrangement is to use as many varieties of common vegetables as you can find, wiring the different shapes in clusters onto a framework of birch twigs.

You will need:

THICK- AND MEDIUM-GAUGE WIRES

BIRCH TWIGS

PLAIN, OVAL-SHAPED MIRROR

GARLIC AND ONION STRINGS

LETTUCES

POTATOES

GLUE GUN

ZUCCHINI

TOMATOES

PERSIMMONS

AVOCADOS

HOPS

1 Wire some birch twigs together to a length equal to the circumference of the mirror. Attach them to the mirror by passing a length of wire through the metal clips, then wind this around the birch twig framework to secure.

2 Cut the garlic and onion strings in half and wire up each grouping. Attach the strings at opposite corners to the birch frame.

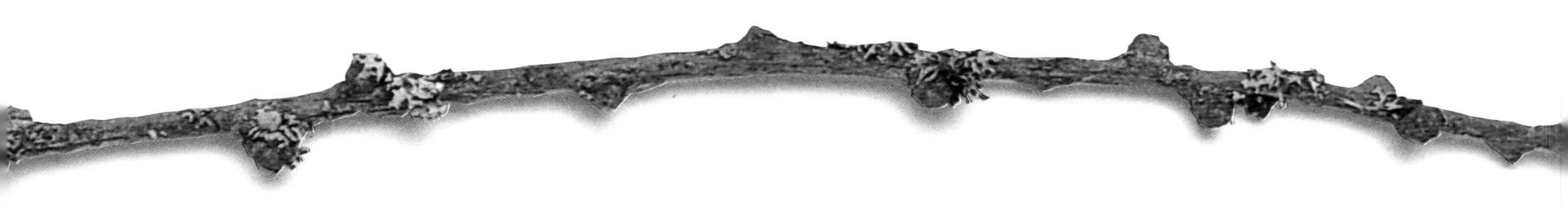

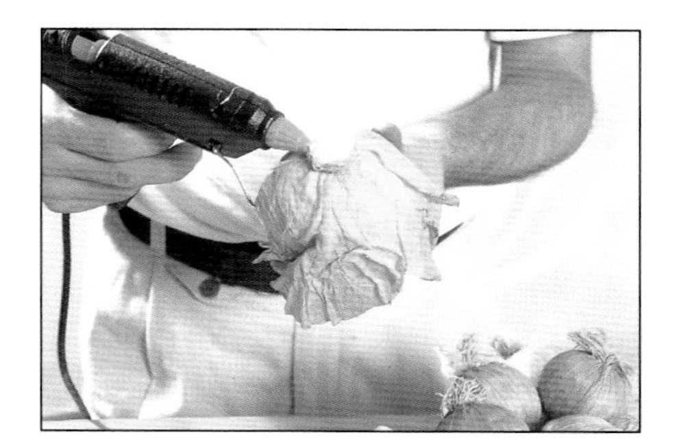

3 Those vegetables which are very light in weight, such as lettuces, can be glued directly onto the birch framework. Use a generous amount of glue applied to the underside of each vegetable.

4 Wire the zucchini, tomatoes, persimmons, avocados, and hops individually. Decide where each grouping will look best on the arrangement and attach accordingly.

5 Build up an even thickness to the garland. When gluing items to the birch framework, hold them down firmly for at least 30 seconds to allow the glue to dry.

6 All the wired fruits and vegetables are flexible and can be adjusted at the end to achieve the best effect. Check that all items are completely secure before hanging the mirror on the wall.

The Cherry Orchard

Create your own attractive fruit basket, then fill it with an abundance of cherries, dotted with a few strawberries, to make a delicious display for a side table or kitchen shelf. The glossy, rich red and burgundy shades of the fruits are perfectly counterbalanced by the soft, fluffy texture of the moss.

You will need:

CHICKEN WIRE

3 LENGTHS OF TWIG

SHARP KNIFE OR SCISSORS

MEDIUM-GAUGE WIRES

CARPET MOSS

GLUE GUN

SHARP KNIFE

FLORAL FOAM BLOCK

CHERRIES

STRAWBERRIES

PLUMS

EXTRA FOLIAGE

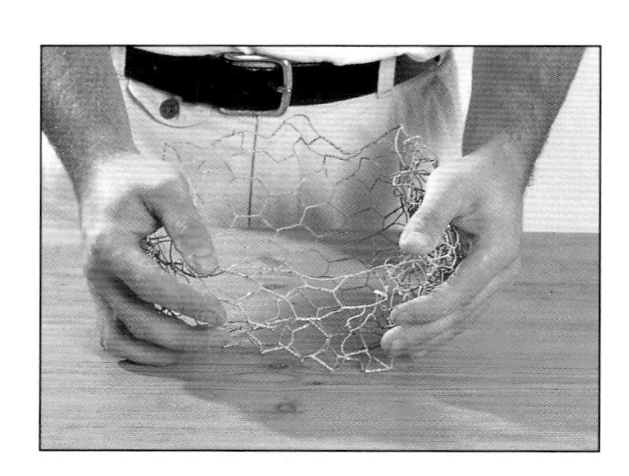

1 Take a length of chicken wire measuring approximately 12 in x 12 in (50 cm x 50 cm) and mold it into a bowl shape.

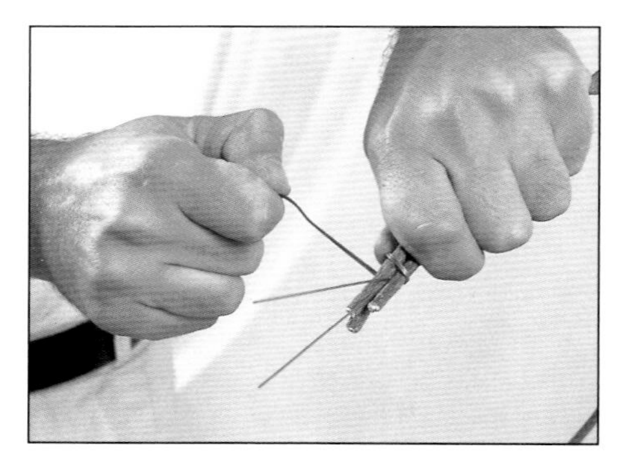

2 Take your twigs and cut them to the same length. Twist the twigs around one another into a rough braid, and bind each end with wire to hold them together. Attach the ends to the top of the chicken-wire bowl to form the basket handle.

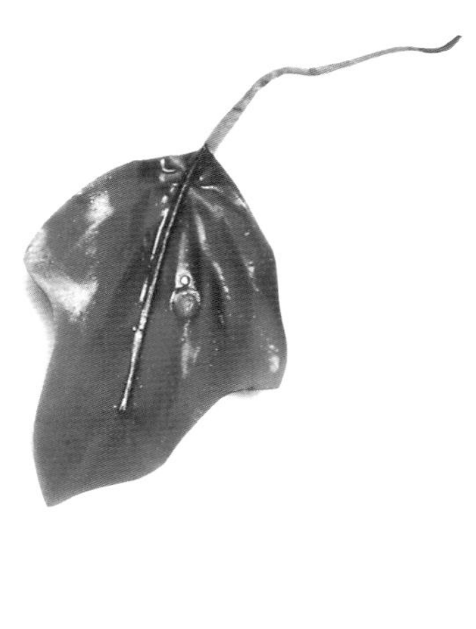

3 Take a clump of carpet moss and place a generous helping of glue on the underside. Press the moss firmly onto the chicken-wire base and hold for a few seconds. Repeat this process until the base is completely covered.

4 Cut a block of floral foam so that it fits snugly inside the moss basket. For added security, the foam can be glued in place at the base.

5 Wire up the cherries in clusters of three. Wire up the plums and strawberries in the same way and begin to insert each cluster of fruit into the floral foam in bold groupings of color. Allow the fruit to hang over the edge of the basket.

6 Check for any gaps between the fruit, where any foam may still be showing. Insert small sprigs of foliage to fill these gaps.

INDEX

Credits

The authors would like to give special thanks to Damian Kelleher, Andrew Moran, Sue Hingston, Mark Todd, Catherine Brading, Monica Maier, and Dee Kendall for their invaluable help during the writing of this book.

Quarto would particularly like to thank **Winward Silks** (see address and details of products below) for so generously providing the majority of the fruits, vegetables, and berries used in this book.

Quarto would also like to acknowledge the following who kindly loaned props for the purposes of photography:

Amazing Grates, London (fireplaces pp45–6, 61–2 & 68–9); Kathleen Barker (champagne glasses pp74–5); Berkertex Brides, Lincs (bridesmaid dress p101); Byron & Byron, London (all curtain rails); Mori Lee UK, Lincs (wedding dress pp9 & 105); Technotile, London (all astro-turf); Wanstead Garden Center, London (wheelbarrow p49).

The fruits, vegetables, berries, foliage, accessories, and equipment used in the projects were supplied by:

Larchfield International
Goodman Street
Leeds LS10 1NP, UK
Tel: 01132 467722
Fax: 01132 467111

Larchfield International have a large selection of artificial fruits, vegetables, and berries.

Winward Silks
30063 Ahern Avenue
Union City
CA 94587
Tel: 510 888 9898
510 487 8686
Fax: 510 886 6888
Toll free: 1 800 888 8898

Winward Silks produce a wide range of artificial fruits, vegetables, berries, and foliage as well as an extensive line of silk flowers. Each item is extremely realistic in terms of color, texture, and weight.